ECONOMIC INCENTIVES FOR ENERGY CONSERVATION

ECONOMIC INCENTIVES FOR ENERGY CONSERVATION

PETER N. NEMETZ
MARILYN HANKEY

University of British Columbia

A Wiley-Interscience Publication
JOHN WILEY & SONS
New York   Chichester   Brisbane   Toronto   Singapore

Library of Congress Cataloging in Publication Data:

Nemetz, Peter N., 1944–
  Economic incentives for energy conservation.

  "A Wiley-Interscience publication."
  Bibliography: p.
  Includes index.
  1. Energy conservation—United States. 2. Public
utilities—United States—Energy consumption.
3. Public utilities—Rates—Government policy—United
States. 4. Energy policy—United States. I. Hankey,
Marilyn. II. Title.

HD9502.U52N4 1984   333.79'17   83-17064
ISBN 0-471-88768-4

Printed in the United States of America

10 9 8 7 6 5 4 3 2 1

PREFACE

The last decade has witnessed a dramatic transformation in the development of modern society. Gone, perhaps forever, is the era of inexpensive energy, the prime mover of our industrial economic systems. On the whole, governmental policy responses to increasing world energy prices and uncertain supply have been weak and diffuse. Of particular note have been regulatory mechanisms which have impeded the efficient allocation processes of the free market. Even in the presence of market failure, where externalities, high transaction costs, and information deficiencies mandate some form of governmental presence, the response has frequently been ill conceived or counterproductive.

Only recently has a more concerted effort been undertaken systematically to apply appropriate economic mechanisms, such as incentives, to the resolution of the significant allocative and distributional questions associated with the production and use of energy.

This book has essentially two goals: first, to examine some recent experience with the innovative use of economic incentives for influencing energy demand; second, to utilize this information to devise functional and effective prescriptions for the development of future energy policy. Much more research remains to be undertaken, as this area entails complex and interdependent economic, political, and social issues. It is our hope that this manuscript will provide an appropriate framework that will facilitate the continuing process of policy formation and execution.

We wish to express our gratitude to numerous individuals for their indispensable assistance in the production of this work: Bert Zethof, for his initial research in the preparation of the Oregon and price elasticity chapters; Alfred Stewart, for his extensive efforts in updating the Oregon case study; Michael Margolick, for his thoughtful comments and contributions to our electricity pricing chapters; Chris Leviczky, for her tireless checking of text, references, and other essential detail; the Word Processors in U.B.C.'s Faculty of Commerce; Mabel Yee; Diane Chajczyk; Sharon Parent; Ulrike Hilborn; Izak Benbasat; and Bill Stanbury and Ilan Vertinsky, for their inspiration, encouragement, and support.

We are especially indebted to those policy analysts and decision makers in California, Oregon, and Wisconsin without whom the production of our case studies would have been impossible. In particular, we would like to thank:

in California,
George Amaroni of the Public Utilities Commission; and Wendell Bakken, Paula Burnette, Commissioner Ronald Doctor, Gordon Gill, Richard Nordahl, Diana Waldie Rains, and John Wilson of the California Energy Commission;

in Oregon,
Sam Campagna, Glen Gillespie, Marsha Henry, Ken Husseman, and Dennis Quinn of Pacific Power & Light Co.; Norm Clark and Steve Hicock of the Bonneville Power Administration; Larry Grey, David Philbrick, William Sanderson, and Tom Wilson of the Department of Energy; Larry L. Payne of the Department of Veterans' Affairs; John Clay and Anthony White of the Public Utilities Commission; John Arthur Wilson of the Northwest Power Planning Council; Carol E. Wisner of the Department of Revenue; Ray Classen, Jeanne McCormick, and Will Miller of the city of Portland; and Marion L. Hemphill, former energy advisor to the city of Portland;

in Wisconsin:
Bonnie Albright of the Department of Administration; Craig Adams and Robin Gates of the State Energy Office; Ken Benkie and James Wise of the Department of Revenue; Professor William Bernhagen of the University of Wisconsin; Benita Byrd, Robert Malko, Gary Mathis, Paul Newman, T. B. Nicolai, Dennis Ray and Mo Reinbergs of the Public Service Commission; James McCambridge of the Department of Industry, Labor, and Human Relations; and Jim Krier, Kathy Lipp and Dennis Hanke of Wisconsin Power and Light Company.

Finally, the authors would like to express their gratitude to the following individuals for their kind advice and assistance in the preparation of this research work:

Jan Acton; John Helliwell; Raymond Hartman; John Anderson of the Minnesota Energy Agency; Lois Arck, Patricia Chapman, Lynn Collins, Al Schwartz, and John Wilman of the U.S. Department of Energy, Washington, D.C.; John Ashworth, Robert Dekiefer, David Roessner, and Mel Simmons of the Solar Energy Research Institute; Eric Hirst of the Oak Ridge National Laboratory; Barbara Kaiway of the B.C. Government Employee Relations Bureau; Larry Kaseman of the Office of Utility Systems, Washington, D.C.; Professor John Miranowski of Iowa State University; and Gordon Pozza of the National Association of Regulatory Utility Commissioners.

Some of this research was originally conducted for Consumer and Corporate Affairs Canada.

PETER N. NEMETZ
MARILYN HANKEY

Vancouver, British Columbia
December 1983

FOREWORD

The period 1973–1983 represents a watershed in North American energy policy and policy analysis. Literally hundreds of measures were implemented by the public and private sectors in response to the events triggered by the Arab oil embargo of 1973–1974. The policies adopted ranged from purely economic ones acting through the pricing mechanism to policies that prohibited energy use for certain purposes. Also, policies that mandated a minimal level of energy efficiency regardless of the cost.

Some of these policies were adopted after careful consideration of alternatives, while others reflected a rush to action in response to public pressure.

Despite the hundreds of millions of dollars invested in these projects, there has been remarkably little analysis of the goals and accomplishments of often disparate policies that were put into effect during this period. Nemetz and Hankey perform a valuable service in drawing across the set of policies that involve cost analysis, pricing policy, investment incentives, and mandatory standards. It is an ambitious charter, and all the more valuable by its very breadth of scope. After all, a policy to reform the structure of electricity rates (e.g., incorporating information about the marginal costs of supply at different times of the day or year) *should* be evaluated in comparison with a policy to encourage more efficient investments in energy-using appliances (e.g., storage electric heaters that could take advantage of electricity produced in less expensive offpeak periods). Yet, remarkably few studies have attempted the comparison.

The evaluation of a diverse set of energy policies includes consideration of economic efficiency, distribution of the consequences across different customers, and the feasibility and acceptability of alternative policies that require governmental, utility, and customer involvement.

Nemetz and Hankey begin their task with a survey of the underlying cost structure of one of the most important sources of energy—the electric supply system. Their choice is a good one because electricity is an important consumer of primary fossil fuels: since almost every business and residential customer uses electricity and electric utility costs and rates are an established matter of public policymaking. Substantial advances in cost analysis and rate setting practices occurred during the 1970s for North American electric utilities. By the early 1980s, almost every electric utility regulatory body had at least begun the process of considering seasonal and hourly variation in average and marginal costs for the major utility systems under their jurisdiction. The authors set the stage by explaining the traditional method of cost analysis, which is bedded in accounting costs, and then move to the marginal cost and peak load considerations—requiring economic and engineering analysis— that constitute the major advance in analysis over this period.

The step from costs to rate policy is an important one involving a number of considerations. Prominent among them are the changes that occur in prices that customers face and the degree of price responsiveness that they display in response to these changes. Nemetz and Hankey draw upon detailed econometric studies from North America and, to a lesser extent, European utilities to identify probable short- and long-run response. They include many first-rate Canadian studies in their review, which enriches the empirical insight considerably.

Having set the background in allocative and efficiency considerations, the authors turn to a review of a whole set of energy policies presented in the form of case studies from states that have been innovative in energy policy matters. To my way of thinking, this is one of the most important contributions of the book. Here the authors juxtapose pricing policies, purchase incentive policies, and policies which mandate certain performance standards. In three case studies they provide a catalog of significant policy developments for California, Oregon, and Wisconsin. They include some of the policies that failed as well as ones that were successful and marshal available evidence of their effects. Many of the policies reviewed in the case studies were good ones—achieving their objectives with relatively low administrative costs and high acceptability by participants; others were notable failures, addressing goals that had never been identified or articulated and carrying significant burdens compared to any benefits achieved. Clearly we need to merge the lessons from this broad experience if we are to make proper use of incentives and performance standards if another energy crisis erupts and policy action is needed in the atmosphere of political crisis.

Nemetz and Hankey conclude on a provocative note and show their interests as policy analysts in the broad sense, not only as economists.

They identify combinations of policies that, taken together, were necessary in order to ensure the success of any one of the policies or in order to enhance the effectiveness of one another. Clearly this is an important direction for further attention in economic and policy analysis, where we often judge policies one at a time. By identifying policies that are synergistic, Nemetz and Hankey not only find several good energy conservation policies, they also contribute to the advancement of policy science.

JAN PAUL ACTON

The Rand Corporation
December 1983

CONTENTS

LIST OF TABLES

LIST OF FIGURES

ECONOMIC INCENTIVES FOR ENERGY CONSERVATION

INTRODUCTION

This book addresses some of the more important issues in the use of economic incentives for the promotion of energy conservation in North America. This work draws on the increasing body of experience in both the United States and Canada and, while reviewing some of the more noteworthy aspects of the achievements to date, focuses on policy prescriptions for future action by both the public and private sectors.

The first five chapters provide an extensive and detailed study of pricing issues in electricity. This form of energy offers one of the most interesting and complex vehicles for the application of economic instruments to the realization of a very important technological and social goal: the achievement of energy conservation with little impact on the health and growth potential of the national economy.

Three case studies follow. These are basically historical in nature and review recent experience in three American states that have pioneered in the application of innovative economic tools to issues of energy policy: California, Oregon, and Wisconsin.

Three appendixes have been provided for interested readers who wish to pursue further some of the more technical issues involved in the use of economic incentives. The first presents evidence concerning demand elasticities for the principal fuels utilized in the residential sector; the second is an illustrative semiquantitative energy response questionnaire; and the third is a compilation and brief assessment of energy conservation measures.

1

ELECTRICITY PRICING: INTRODUCTION AND THE NATURE OF UTILITY COSTS

1.1 ECONOMIC AND POLITICAL BACKGROUND OF RATE REFORM

Public attention was focused on utility rates after the oil crisis in 1973 forced the first of a series of rate increases. The rising fuel costs underlying the rate changes were the most visible symptoms of a broader disorder affecting the industry. Increasing costs of capital, higher construction costs, pollution control equipment, and a slowdown in the rate of technological improvements all contributed to the need for higher rates. Under regulatory scrutiny, utility problems became more clearly defined and means were actively sought to remedy these problems. When it became evident that utilities no longer faced increasing returns to scale but rather faced rising costs with increased output, the debate turned toward reform of the rate structure to match the changed circumstances. A truly cost-based rate (marginal cost pricing) or intensified use of existing baseload capacity (load management) were claimed by their respective proponents as the best means of ultimately lowering electricity costs.

The utilities, faced with a deteriorating capital position and diminished investor confidence, responded by deferring plant additions or attempting to improve system reliability. One response has been increased interest in the interconnection of systems. The trend of declining load factors (the ratio of average utility load to peak load), down from 65% to 61% over the period 1965–1975 in the United States, helped make the utilities receptive to suggested reforms (U.S. FEA, 1977a, p. 4).

Once the debate took shape as a political issue, interest groups emerged, recognizing that major changes in pricing were at hand. Industry and consumer groups were vigorously opposed to higher rates, which conservation and environmental groups supported as a curb on consumption. Manufacturers of meters, computers, and load management devices encouraged the shift to marginal cost pricing.

Theoretically, marginal cost pricing is a neutral and economically defensible concept. It is defined simply as the cost of producing one additional unit of output. As will become evident, economic theory does not specify which increment of output is to be measured, nor does it provide any practical means of actually measuring costs, once it has been decided which costs are to be measured. Furthermore, in competitive economic theory, certain unrealistic assumptions are present that do not hold in the real world, such as perfectly divisible inputs, the absence of risk and uncertainty, and the pricing of all other goods in the economy at marginal costs (CEC, 1979, p. 8). The nature of electricity generation itself is peculiarly resistant to these assumptions. Output, and to some extent input, is available in blocks rather than incremental units; risk is a measure of system reliability; and no one can pretend that other goods, particularly energy resources, are priced at marginal cost. The argument most frequently voiced against marginal cost pricing is that it is an all-or-nothing proposition; that is, it cannot be applied piecemeal within one sector of the energy industry. One report found that there is

> no evidence in the record to support the proposition that a little bit of marginal cost pricing is better than accounting cost pricing. . . . In effect, marginal cost pricing became little more than a catchword, without completeness, stability or consistency. There is no clear, practical definition of marginal costs and no clear, practical way of reconciling these costs with the revenue requirement. These obstacles are insurmountable (O.E.B., 1979, pp. 53–54).

The California Energy Commission also found all prevailing methodologies for calculating marginal costs unacceptable and has attempted to derive a new method. But the resiliency of the concept of marginal costing and the ingenious assortment of costing methodologies attest to the fact that it will not fail to influence electricity pricing in the future. Given the monopolistic structure of utilities, where no market pressures exist to push rates to an efficient structure, some form of marginal cost pricing must be imposed to encourage economic efficiency.

Major issues in electricity policy include the "role of nuclear power, and the internal inconsistencies of the pricing, demand forecasting and investment planning of electric utilities" (Helliwell, 1979, p. 177). The choice of a satisfactory pricing mechanism influences demand, which in turn influences the scope of energy supply options. Overprojecting demand for electricity (based on continuation of declining-block rates) and the maintenance of ultrasecure reliability levels contribute to expanded capacity, increased capital requirements, and higher rates.

By way of example, several major utility systems in Canada (such as Ontario Hydro) have considerable excess capacity. In Ontario, energy demand and supply studies conducted in 1975 assumed that the provincial system was optimal, and there were even warnings of potential outages. Utility forecast growth was put at 7% per annum. By 1978–1979, the forecast of annual growth had been reduced to 4.7% through 1990 and to 4.2% thereafter to the year 2000. A more recent report concedes that even these figures might be too high. The overall system excess reserve capacity was estimated at 45% (15% is considered normal), and annual rate increases, which reached a high of 30.3% in 1977, declined to 9.8% in 1979 (O.E.B., 1979, pp. 7–8). These facts defused the sense of urgency relating to rate reform and undoubtedly helped the resistance to the adoption of new pricing principles.

The role that a government assumes in electricity pricing will be determined by its position on other energy issues—the relationship of energy prices to social and opportunity costs, energy conservation, the scale of energy projects, commitment to alternative technologies, and the income distribution effects of rising energy prices. A position on each of these issues can be reflected in the price of electricity. Because of this, and because it touches each household directly, the price of electricity is a powerful policy instrument.

1.2 THE NATURE OF UTILITY COSTS

1.2.1 Generation Costs

Cost classification in the utility industry is based on service characteristics such as demand, energy consumption, and the number of customers served. Demand costs are defined by rate engineers as those costs incurred to meet peak demand, where "demand" is the maximum output of electricity at any point in time measured in kilowatts (kW). Demand costs are those cost components reflecting fixed investment in generation, transmission, and distribution systems and usually include a carrying charge on the investment in plant and equipment. Energy costs are

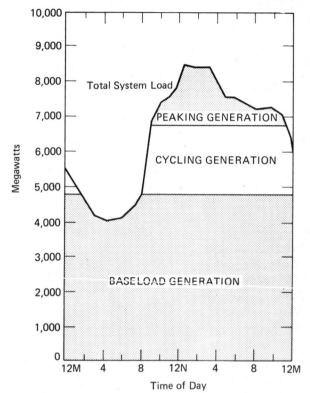

FIGURE 1.1. Dispatching generation to meet a cyclical load. (Source: NARUC, 1977, p. 15; © 1977 Electric Power Research Institute. Reprinted with permission.)

those operating expenses varying with energy consumption measured in kilowatt-hours (kWh). In addition to fuel costs, variable operating and maintenance expenses are classified as energy costs.

Traditionally, the capital cost of generating capacity, which includes a mixture of base, intermediate, and peaking plants, is considered a fixed cost causally related to system peak demand. Demand and capacity relationships are graphically represented via load curves and load duration curves. Figures 1.1 and 1.2 illustrate a daily load curve; Figure 1.3 is an annual load duration curve. System peak demand and a safety margin (15–25%) determine the necessary amount of generating capacity. Peaking capacity is required to meet maximum demand, although it remains idle most of the time. Baseload generation capacity is capital intensive with low running costs, while peaking capacity is expensive to operate but relatively inexpensive to install. Therefore, the cost of providing electricity to customers varies widely, depending on the time of day, day of the week, and season. For each plant total costs rise as the number of

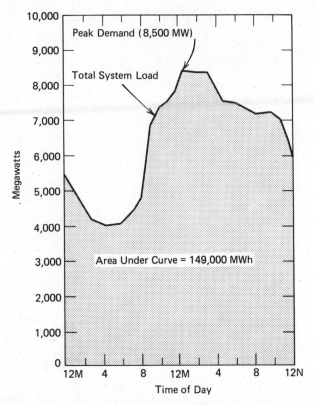

FIGURE 1.2. Daily load curve. (Source: NARUC, 1977, p. 15; © 1977 Electric Power Research Institute. Reprinted with permission.)

$$\text{Daily load factor} = \frac{\text{daily energy}}{24\,\text{h} \times \text{peak load}} = \frac{149{,}000\,\text{MWh}}{24\,\text{h} \times 8500\,\text{MW}} = 0.73 = 73\%$$

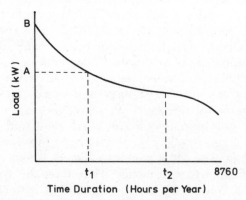

FIGURE 1.3. Hypothetical annual load duration curve. (Source: U.S. FEA, 1977a, p. 31.)

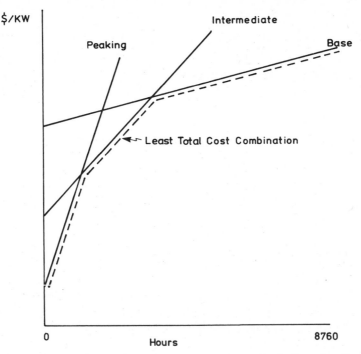

FIGURE 1.4. Screening curve analysis. (Note: Generation is divided into baseload, cycling [intermediate, mid-range] and peaking generation. Baseload capacity is the most capital intensive and tends to be the most efficient. Peaking units are the least capital intensive and have the highest fuel costs. Cycling units are in between and often coal fired. This figure illustrates the relative type of cost for each type of capacity.) (Source: CEC, 1979, p. 55.)

hours of usage increases. After a certain point, additional operation of one unit will result in a higher total cost than if some other unit were used. The planner tries to fill the area under the load curve with the least cost combination. A planner will choose to employ baseload capacity as long as its total running cost is less than or equal to that of an intermediate (cycling) unit. He will do the same with cycling unit use before employing peaking units.

An important distinction must be made between long-term planning criteria (where future capital costs are flexible) and short-term operating decisions which are constrained by implaced, "sunk" capital. In the latter circumstance, fuel costs are the principal consideration in system output decisions.

Figure 1.4 illustrates the least cost combination of equipment from a longer-term perspective. The slope of the three curves represents the relative costs of each type of capacity.

The effect of successful load management policies is to flatten the load curve (equivalent to rotating the load duration curve counterclockwise). This can be done by either reducing the peak or increasing the baseload. Because electricity is not a homogeneous commodity that can be stored easily, its cost varies with the time generated. The idea of marginal cost pricing of electricity is therefore tied to generation (and, to a lesser extent, to transmission and distribution) by time of day.

If the fuel used in peaking units is determined to be the basis of marginal costing of peak-period generation, time-differentiated rates based on this principle will vary widely between installations. Fuel costs as a proportion of total operations and maintenance costs in Canada range from over 75% in Prince Edward Island to less than 1% in Quebec. Quebec, Newfoundland, Manitoba, British Columbia, and the interconnected systems of the Yukon and Northwest Territories all have over 90% of their electrical energy generated from hydro sources (EM&R, 1980, p. 71). In these provinces peak-period marginal costs defined as fuel-related only would be a small fraction of total operating costs; consequently, the cost differential between peak and off-peak periods would be small. The rest of the provinces are more diversified in their generation mix, and the trend is toward increasing diversification. In Ontario, hydraulic generating units operating as base, intermediate, and peaking plants account for approximately 34% of total electrical energy output; nuclear-powered base plants provide about 27%, and the balance is provided by fossil fuel thermal plants (29%) and purchases (10%) (O.E.B., 1979, p. 109).

The method by which fuel is costed depends on the type of fuel and the accounting principle employed. Ontario Hydro charges fuel costs to the cost of power on an average-inventory-price basis when the fuel is consumed. Coal costs are calculated monthly on a rolled-in basis, including rail and barge freight, inspection, and storage charges. Oil too is priced on an average basis; consumption of natural gas is costed at the contract price. The cost of steam purchased from Atomic Energy of Canada Ltd. is treated as a fuel cost, as is power purchased from other utilities (Ontario Hydro, 1976b, p. 13). Each accounting method will yield a different cost. For example, average costing smooths cost differences and tends to collect from current customers some of the funds required for inventory replacement. First-in-first-out and last-in-first-out (not accepted for tax purposes) will lower and raise fuel accounting costs, respectively, as prices rise.

The conclusion to be drawn from this look at fuel costs is that even the simplest definition of marginal operating costs (i.e., fuel expenses) is highly arbitrary in practice. The issue becomes more clouded when the debate extends to whether or not, and in what proportions, baseload capacity should be included in peak-period generation costs.

1.2.2 Transmission and Distribution Costs

There is no clear line of demarcation between transmission and distribution (T&D) other than an arbitrary voltage level below which one decides that the distribution system begins (usually 115 kV). Transmission costs involve those expenses incurred in the transmission of power from generation facilities to the distribution system. Because transmission lines are designed to carry a specified maximum load, costs associated with transmission are demand (kW) related. Line losses are somewhat greater during peak periods as well, so transmission costs can be said to vary with time of day. Transmission costs account for about 9% of total system costs (U.S. FEA, 1977a, p. 37).

The costs of distribution are related to the low-voltage lines that transport power from the transmission grid to the customers' supply points. These lines are included in the cost function on the basis of kilowatt-miles. The other fixed components of distribution costs—cables, transformers, circuit breakers, poles, and meters—are classified as customer-related expenses. The demand component of a distribution system recognizes the load requirement and the losses resulting when line capacity is approached.

Distribution costs account for approximately 18% of total system costs (U.S. FEA, 1977a, p. 37) and vary with the number of customers served. Industrial customers who are able to take their power directly from the transmission grid pay no distribution costs.

In defining marginal costs in transmission and distribution systems, the issue is refining the distinction between the customer and demand-related components. Very little attention has been paid to this problem, although T&D systems represent a significant percentage of the total system costs. The European practice is to analyze marginal generation and transmission costs at the highest voltages, calculate rates based on time-of-use patterns, and increase these rates by the costs of distributing power at successively lower voltages (Mitchell and Acton, 1977).

1.2.3 Customer Costs and Metering

Customer-related costs are a direct function of the number of customers on the system. Conceptually, these costs would be incurred whatever amount of electricity was being consumed. They constitute expenses for billing, metering, connection, administration and overhead, and represent a very small proportion of total utility expenses.

1.2.4 The Relationship Between Load Factors and Cost

The "load factor" indicates the degree of utilization of capacity or variation in demand. Numerically, it is the ratio of average demand to peak demand. Because peak demand (maximum kilowatt demand) determines the capacity of the system, peak demand also determines the necessary size of capacity and hence fixed costs.

Changes in load factors affect production costs in a number of ways. An increase in this factor with no change in kilowatt-hours sold should never raise any of total costs or short- or long-run marginal costs of a utility. If the baseload plants are not being run at full capacity, such an increase in load factor will lower total costs. The transfer of a given kilowatt-hour load from peaking to baseload units leaves per kilowatt-hour costs of running the peaking plants unaffected while reducing the per kilowatt-hour costs of baseload electricity production. Therefore, an increase in load factor with no change in kilowatt-hours sold with baseload plants not fully loaded should reduce total costs.

The same argument applies to short-run marginal costs. When baseload plants are not fully loaded, the shift of a fixed kilowatt-hour load from peak to baseload raises the short-run marginal costs of the baseload facilities by less than it reduces the fuel costs of the peaking plants.

Long-run marginal cost should also never be increased by load leveling. The effect of leveling will generally be to accelerate baseload plant construction, while deferring or reducing use of peaking plants. A priori one might think that this strategy would always lower total costs because baseload plants are cheaper per kilowatt-hour to run (in combined fuel and capital cost). However, if the utility would have to build a very expensive, large baseload plant to accommodate a perceived shift from peaking to baseload demand, it would be worthwhile to avoid new baseload construction and simply generate the same number of kilowatt-hours in each class of plant as before. Long-run marginal cost would be reduced to the extent that new peaking units would be deferred. Baseload plant construction scheduling would remain unaffected.

Logically, costs per kilowatt-hour should fall as the load is flattened because expensive peaking units are not being employed. But this again relates to how those costs are defined and to how overall consumption responds when the peaks are reduced. The response in consumption will be influenced, if not determined, by the rate structure in effect. The commonly employed two-part rate consists of a demand charge based on that customer's maximum peak demand over some period of time (e.g., 15 minutes in a month). That customer's peak may or may not coincide with system peak. When it does, it is known as "coincident peak." A peak charge that is noncoincident therefore bears little relationship to the costs that a customer imposes on a system. In most formulations of marginal cost pricing, the demand (kW) charge (which distributes capacity costs) is reduced in relation to energy (kWh) charges. This may actually reduce load factor improvement incentives (Carpenter, 1978, p. 28).

Canadian system load factors may not be deteriorating in the way observed in U.S. systems. The average Canadian annual load factor has stayed between 65.5% (1974) and 71.3% (1976) since 1960. Most Canadian utilities report that daily load factors have been improving, except Quebec and to a lesser extent Manitoba, which have been experiencing deteriorating load factors. The daily winter utilization factor for large networks is a healthy 85–90% (Protti, 1978, pp. 270–271). Utilities with very high load factors display little variation in hourly costs.

1.2.5 Social and Opportunity Costs

In a truly comprehensive energy pricing system, the formula for the price of electrical energy (P_e) could be expressed as $P_e = MC_p + MC_{it} + MC_e$, where MC_p is the private marginal cost, MC_{it} is the intergenerational transfer marginal cost, and MC_e is the environmental marginal cost (Ontario Hydro, 1976f, p. 8). Conservation of energy, in economic terms, is a redistribution of resource use to the future. (The opposite, depletion, is a redistribution of resource use to the present.) If the term MC_{it} is positive, this would imply that current users of the resource in question are imposing a cost on would-be future users, and this cost should be reflected in the current price. Therefore, MC_{it} is the opportunity cost in terms of alternative future uses of present electricity production (Ontario Hydro, 1976f, p. 8).

The use of world prices represents the social marginal, or opportunity cost of oil. It is the price that could be realized in the market if the oil were not domestically utilized, or the price a utility would have to pay for an additional barrel of oil without the benefit of any implicit or explicit subsidies.

Environmental costs could be accounted for by adding the value of any derivative harm inflicted on the environment to the private marginal cost. If the costs of these negative environmental effects ("negative externalities"), which could include even aesthetic losses from ill-placed structures and lines, were added to the price of a kilowatt-hour as a tax, excess revenues would inevitably result. The problem of excess revenues arises whenever full marginal costing is debated because of the extent to which these real, albeit elusive, social costs have escaped quantification and inclusion in the price structure. Some have argued that the use of "shadow pricing, or an externality tax, should be rejected as an alternative for pricing policy. [A utility] . . . should base its prices on marginal private costs, while meeting the prescribed environmental standards set by the government" (Ontario Hydro, 1976f, p. 21).

Other kinds of direct expenditures are not included in the price of electricity because they are made by the government rather than utilities. A large proportion of both Canadian and American federal energy research and development expenditures is related to nuclear energy. It is evident that the total costs of nuclear power are significant; yet, because of invisible subsidies, the per-kilowatt-hour generation costs of a nuclear-powered plant are frequently thought to be among the least expensive. Any reformation of electricity rates that entails reductions in the expected growth of electricity demand will substantially reduce the planned program of nuclear power development (Helliwell, 1979, p. 215).

1.2.6   Summary and Conclusions

The classification of a utility's fixed and variable costs is a matter of judgment. The costs are affected by the accounting principles applied, as well as the broader principles of welfare economics, which attempt to reflect social and opportunity costs. Load factors appear not to have deteriorated as markedly in Canada as in the United States. Improved load factors with no overall decrease in electricity consumption may result in increasing capital expenditures for more baseload generation. Fuel costs should represent the value of that fuel in its best alternative use. Risks, alternative uses, and environmental effects should be defined more clearly and included in marginal cost calculations.

2

ELECTRICITY PRICING: COST CLASSIFICATION AND RATE-MAKING CONSTRAINTS

2.1 COST CLASSIFICATION AND BEHAVIOR

2.1.1 Accounting Costs

The terms *accounting costs, average costs,* and *embedded costs* are used roughly interchangeably in the utility industry. Depending on the base established for the revenue requirement—current, past, or future year—embedded costs are the costs recorded on the books of account, including depreciation expenses, which theoretically cover the loss in value of existing plant and the fund from which new capacity is to be financed.

The two main criticisms leveled against the usual methods of rate base accounting involve the implicit use of averaging and the write-off of plant and equipment.

Three basic types of averaging of costs occur in a fully distributed cost system: (1) *historical averaging,* that is, the costs associated with older plants are averaged in with current costs of new plants; (2) *plant and energy cost averaging,* where the technology mix of all generating plants is averaged to determine the demand (kW) charge, and the energy (kWh) charge represents an average of the variable costs associated with each type of generation; and (3) *cost averaging through time,* where the higher costs of providing peak power (by season and day) are averaged in with the lower costs of providing energy during the off-peak periods. The use of averages will generally understate the cost that a utility incurs in supplying additional electricity when there is real cost escalation. Because the demand for electricity is sensitive to price, the result will be an overbuilding of plant to meet excess demand (Ontario Hydro, 1976f, p. 4).

The use of straight-line depreciation implies that the current value of existing assets is declining. In times of rising prices and slowing technological improvement, the economic value of a piece of equipment should actually increase as the cost of replacement rises. Average costs calculated with continually depreciated assets and low debt service obligations keep rates lower than warranted by the fact of inflation. Marginal costs will always exceed revenue requirements based on historical costs under these conditions because they are forward-looking and not based on historical costs. Only if fuel prices and technology are stable will rates based on marginal cost be comparable to those based on embedded costs, that is, past and future costs will be equal.

Time-Differentiated Versus Non-Time-Differentiated Accounting Costs

Marginal cost is commonly used as an optimum measure of output in economics. A competitive firm produces at an optimum level when the cost of a unit of output equals its market price. Because price is the best measure of the value society places on output, it should be matched to marginal cost, the best measure of real social cost of output. When marginal cost equals price, costs are matched equally with benefits, and the welfare of society is maximized. If the benefit price (or marginal revenue) is greater than marginal cost, production should be expanded. If marginal cost is greater than the selling price, resources are being wasted increasingly as production expands.

Marginal cost is by its nature a forward-looking concept. It is the cost of the *next* unit of output; or the cost of using *more* electricity. In power generation this can be the cost of either the next kilowatt or kilowatt-hour; that is, it can be used to account for both the demand (kW) and energy (kWh) costs. The marginal cost of a kilowatt-hour consists primarily of the fuel cost of the most expensive generating unit used to provide the additional kilowatt-hour of power.

There are few who dispute the idea that marginal costs vary with time of use. The question is whether those who create the additional demand are also responsible for the underlying baseload generation costs because of the fact that system generation capacity is designed for maximum demand. Off-peak increases in demand can be matched to the variable costs of baseload capacity. And there is a general consensus that the marginal costs associated with an increase in peak demand should include the operating and capital costs of peaking capacity. But even if one could identify exactly who is responsible for the system peak, it may be hard to argue that his consumption, and not another's, is more responsible for peak capacity costs.

The level of marginal costs is influenced by both supply and demand factors. On the supply side, the type of generating mix, choice of accounting convention, and costs of capital and fuel all contribute to determining a level of costs. On the demand side, climate is perhaps the biggest determinant of electricity demand, as well as the cost of competing fuels and the level and type of industrial activity. Because each of these factors differs between utilities, and certainly between countries, direct cost comparisons between utilities have little meaning.

Comparison of marginal costs with embedded costs produces profound discrepancies which are at the root of the debate. First, marginal cost considers the cost of new capacity, which will invariably be higher than the cost of existing capacity. In a marginal cost analysis, previously incurred expenses are considered sunk costs and are therefore irrelevant to rate setting. Second, most methodologies for calculating marginal cost are based on the concept of providing increments of power on a least-cost basis, whereas traditional accounting methods average capacity costs (U.S. DOE, 1979d, pp. II-5–II-6). Consideration of future capacity costs will yield marginal costs in excess of embedded costs. Nevertheless, it cannot be stated with certainty that marginal costing will always yield excess revenues, particularly if rates based on marginal costs cut total utility costs by altering the pattern of demand. One advantage of marginal costing is that peak-period rates should generate revenues matched to actual costs over time. At present, every incremental unit of demand is being sold at less than it costs to produce, so that increases in demand compound losses. The "price signals" spoken of in conjunction with marginal cost pricing reflect the costs of increased consumption related to plant expansion or high-cost generation fuels. Prices based on historical or average costs understate the consequences of users' behavior.

A basic principle of both Canadian and American oil and gas pricing policy is to allow significant increases in previously controlled domestic energy prices. In the United States this involves the "deregulation" of both oil and gas, whereas in Canada the increase of energy prices toward world levels remains a regulated process. An additional feature of the Canadian energy pricing system is a conscious underpricing of natural gas in relationship to oil in order to promote interfuel substitution to a more abundant fuel. Because most of the marginal costs associated with peak-period generation are fuel costs, electricity should be priced on a comparable basis with alternative fuels, if not throughout the rate structure at least during peak periods.

The relationship between the price of electricity and the price of alternative fuels is a difficult problem that has no easy solution in theory or practice. The precondition that all goods be priced at marginal costs for maximum social welfare was a problem that economists escaped by positing the "second best solution," that is, the next best possible solution. If the costs of other fuels diverge from marginal costs, it can then be argued that the cost of electricity too can be priced on a nonmarginal basis. Inasmuch as fuels are close substitutes, any pricing scheme that favored electricity consumption at the expense of other fuels would eventually drive up electricity prices. In the meantime, increased consumption of electricity would accelerate the utilities' losses per unit of output (Wisconsin PSC, 1979f, p. 15).

2.1.2 Marginal Costs

Long-Run Marginal Costs

Marginal costs are not calculable without reference to a future time horizon. Long-run marginal costs (LRMC) are related to future alterations in capacity to meet a given increment of demand.

One practical method of calculating long-run marginal costs is to compute the present value of revenue requirements over a long period associated with a given expansion plan (generally 10–15 years) based on a set of demand expectations. By varying these demand projections and recalculating revenue requirements, one can estimate production level LRMC. The marginal cost of peak demand would be calculated as the least cost method of meeting a uniform increase in kilowatt peak demand over the planning horizon. There are many methods for calculating LRMC, but each must be capable of estimating costs for changing circumstances.

Long-run generation costs are calculated relative to peak demands. The LRMC in the peak period is either the cost of constructing and operating a new peaking unit or the cost of a new baseload unit less fuel savings. Transmission and distribution costs are allocated to periods, as are generation costs. One method of allocating costs to time periods is via loss-of-load probability (LOLP)—the probability that a shortage of capacity might be experienced in any given period. Alternatively, costs could be allocated to rating periods on the basis of relative energy costs; both this measure and LOLP increase as load levels increase. Demand-related distribution costs would be based on projected growth in the number of customers only.

TABLE 2.1.   A Quick Guide to LRIC and Time-Differentiated Marginal Costing

Costing	LRIC	Time-Differentiated Marginal Costs
Capital cost of generation	Mean expected plant cost per kW over planning horizon. Current dollars.	Least capital intensive plant used on system, usually a peaker. Current dollars.
Annualization (carrying charges)	Levelized at current rates of interest, including taxes.	Annual charge reduced by (approximately) rate of inflation, includes taxes.
Fuel costs	Average fuel cost. Current prices.	Marginal fuel cost (always > average). Current prices.
Distribution and transmission	Mean per kW cost of recently installed capacity.	Mean per kW cost of recently installed capacity.
Customer costs	(Judgmental)	Minimal system.
Capacity responsibility	Some measure of peak responsibility.	Loss-of-load probability.

Source: Ontario Hydro, 1976g, p. 8.

Contrary to most studies, an Ontario report concluded that the use of LRMC was not supported by economic theory (O.E.B., 1979, p. 44). The concept of LRMC was rejected because no consistent definition could be found and because conflicting definitions would yield significant differences in practice. This criticism has led to the development of a substitute measure of LRMC called long-run incremental costs (LRIC). The differences between the two pricing systems are set forth in Table 2.1.

The use of the LRIC system is similar to future-test year methods for establishing revenue requirements; it is based on the future financial profile and cash flows of the utility and is therefore more consistent with current practices. It projects capital and operating costs into the future and estimates the incremental cost of meeting additional demand. In contrast, long-run marginal costs are not tied to cash flows, but rather seek to measure the future costs of consumption on a time-differentiated basis. Calculation of LRMC involves not a single method but a variety of available techniques.

The choice of a discount rate reflecting expected levels of inflation as well as technological progress exerts a powerful influence on the results obtained in LRMC calculations. *Expected* future technological progress tends to "raise" *present* marginal costs (technological progress is accounted for by raising the discount rate used), since by building today we lose the opportunity to take advantage of tomorrow's innovations. Similarly, if the rate of interest used to discount future costs overestimates the expected inflation rate, lower marginal generating cost estimates will result, and vice versa. A related issue is whether or not construction in progress should be included in the calculations.

One reason to set rates equal to long-run marginal costs is that customers are then better able to make long-run consumption decisions. If no decision is made to base rates on LRMC, some means should be devised to provide the public with a prognosis for future rates. This has been suggested by Vickrey [1978]. The anticipated adjustment of oil and gas prices to world levels would also encourage energy consumers to use world prices when designing new plants and homes. The continuing uncertainty about future energy price levels makes long-run marginal cost pricing even more troublesome.

Short-Run Marginal Costs

Short-run marginal cost is the cost of providing an increment of electri city when the level of capital stock is fixed, usually defined over a period of one year. When measured in current dollars, short-run marginal costs (SRMC) are usually greater than or equal to long-run marginal costs because a utility increases output by running existing equipment longer or utilizing temporary, high-cost capacity. In the long run, this costly peaking capacity can be replaced with lower-cost, more efficient capacity.

Transmission and distribution costs are generally omitted from short-run marginal cost calculations because they represent fixed capital investment that does not change in the short run. This is an important and justified omission because capital invested in transmission and distribution facilities may be as large as that invested in production plant (U.S. DOE, 1979d, p. J-2).

In practice, the calculation of SRMC begins by defining rating periods, either by season or time of day. Once the periods have been chosen, there are three alternative methods of allocating costs to periods:

USE OF SYSTEM LAMBDA

The use of the system lambda, or hourly running rates, as a basis for cost calculations requires data on period costs and energy losses by voltage and load level. This information is maintained by most utilities. Generally, the short-run response to changes in peak load is equivalent to accelerating or decelerating combustion turbine generators, so the short-run incremental costs amount to the variable costs on a combustion turbine. When calculated, these costs can be allocated to periods using relative loss-of-load probabilities.

USE OF SYSTEM PLANNING MODELS

Another method calculates increased costs relating to a given increase in load for each predetermined rating period. For example, for a base running period, one calculates the fuel costs, then recalculates assuming an x increase in load per year over the period. The difference in the two costs divided by x represents the SRMC for that rating period. Costs are again allocated to periods using loss-of-load probabilities or some other method.

USE OF LINEAR PROGRAMMING

Linear programming models attempt to maximize or minimize a specific function subject to a set of constraints. To calculate SRMC, a model might be used that determines minimum costs involved in meeting an increment of demand, with the added constraint that capital stock is held constant. The shadow prices derived from the solution indicate the marginal costs at specific load levels. All three methods ignore the issue of whether the existing system is optimal (U.S. DOE, 1979d, p. J-7).

Short-run marginal costs tend to be volatile because of frequent changes in supply and demand conditions. Compared to LRMC, they lack the stability that enables consumers to make long-run investment or consumption decisions. Because the type of plant one employs is allowed to change under the assumptions of LRMC, an oil-fired plant could be replaced by a nuclear plant in the long run, which would give very different price signals to consumers. A production mix that was optimal during an era of low fuel prices is now extremely expensive to run. At present, rates based on SRMC would be greater than those based on LRMC. The emphasis of SRMC lies with fuel costs and the value of reduced system reliability (estimated costs of outage multiplied by the probability of its occurrence) rather than the effects of demand on future capital expenditures.

2.1.3 Summary and Conclusions

The difference between using accounting and using marginal costs has been well summarized by Turvey (1978, p. 2): "Marginal cost pricing provides relevant though uncertain information about the future to electricity customers. Embedded cost pricing on the other hand provides irrelevant but certain information." Non marginal, cost-based rates camouflage the costs to society of using more electricity. Rates based on average, historical costs presently lie below marginal costs and imply a loss incurred for each additional unit sold. All formulations of marginal cost tie costs to time of use, although time-of-use rates can be made consistent with traditional accounting methods. Long-run marginal costs (LRMC) include plant additions, whereas short-run marginal costs assume fixed capacity. Long-run incremental costs (LRIC) are a proxy for LRMC and are related to current accounting practices. Short-run marginal costs do not project the consequences of consumers' consumption decisions into the future as do LRMC, but instead reflect immediate changes in system costs.

2.2   RATE-MAKING CONSTRAINTS

2.2.1   Investment Patterns

The major constraints facing a utility selecting a capacity expansion plan are the following:

Availability of capital.
Fuel availability.
Environmental considerations.
Supply agreements with other utilities and power-sharing pools.
Maintenance scheduling.
Nuclear refueling.
Construction and regulatory lead time.

To this list might be added the impact of load management strategies if adopted.

Utility planning is an effort to match system expansion with forecasted demand growth. The rate of expansion, and therefore the necessary capital requirements and rates, reflects estimates of growth. Until recently, utilities conducted their own forecasts. The demand forecasts used by Canadian utilities in their 1975 expansion plans, surveyed by Energy, Mines and Resources Canada (EMR), were found to be so high as to be inconsistent with any of EMR's econometrically derived forecasts (Helliwell, 1979, p. 211). The EMR load growth forecast (under assumptions of high energy prices) of 538 billion kWh by 1990 was 17% below that of the utilities.

The EMR forecast was criticized on the grounds that its projections involved unrealistically high energy savings. [Price-responsive demand models tend to overpredict the speed of response to large and rapid energy price increases (Helliwell, 1979, fn. 42).] Yet when the published EMR model was applied to 1976 data, its forecast was still 5.4% above the actual sales volume for that year.

Assumptions of slightly greater industrial price elasticity of demand would make the EMR model-based forecast of 1990 electricity sales 30% less than that forecasted by the utilities (Helliwell, 1979, p. 213).

Statistics Canada predicted that electric power demand would grow at a compounded annual rate of 5.25% over the period 1978–1983. This estimate is an average encompassing a low 4% forecast growth rate for Ontario to a 9% rate in Alberta. Substitution of electricity for heating oil contributed to higher electricity growth rates in the Atlantic provinces and Quebec, and part of the increased demand for electricity in Alberta is attributed to the power needed for oil sands production (*The Financial Post*, 1980, p. 29). Past Canadian growth in electricity demand had been averaging 6.8% per annum since the mid-1950s.

A more important indicator of the relationship between demand forecasting and expansion is evident from the levels of reserve capacity maintained by Canadian utilities. It has been projected that reserve margins measured as a percentage of peak load will increase in Ontario to 47% in 1983 from 44% in the late 1970s; at New Brunswick Power the margin will increase to 32% from 18%; in Saskatchewan the margin will nearly double to a projected 27%; B.C. Hydro will maintain reserves at 36–38%; Nova Scotia and Manitoba will lose 3–4% reserve capacity as their margins drop to 49% and 37%, respectively; and Quebec Hydro will reduce its margin of 12% to just over 7% by 1983 (*The Financial Post*, 1980, p. 29).

Forecasting is undoubtedly an imprecise science. When demand is overestimated, prices tend to be set to encourage expansion of demand to meet the level of prebuilt capacity. In the view of John Helliwell (1979, pp. 213–14), this tendency to overbuild perpetuates itself: High average prices are necessary to cover the costs of overbuilding, and these, in turn, restrict demand and delay the rate structure reforms necessary to align marginal costs with revenue. Helliwell feels that the financial pressure exerted on utilities should motivate them to use rate structure reform to cut the rate of expenditure growth. Until that time, however, excess electricity supply conditions in Canada will continue with capital and operating costs at unnecessarily high levels.

The tendency to overbuild has been analyzed by economists in the context of utility regulation. The 1962 study by Averch and Johnson found that regulated rates of return on capital, if more than the marginal cost of funds to the firm, would cause utilities to use excessive amounts of capital relative to the amounts necessary to minimize total costs for a given output (Helliwell, 1978, p. 115). Inducing utilities to adopt cost-minimizing behavior is a difficult regulatory objective. If prices are to be set at marginal costs, some assurance must be offered that those costs are at a minimum.

2.2.2 Revenue Requirements

One of the major objections to initiation of marginal cost pricing is that its revenue yields are uncertain. Conversely, because average-cost pricing is based on actual book expenditures over the term of the applicable rates, revenue requirements under this traditional pricing system can be matched fairly precisely. The primary emphasis in marginal costs is determination of an appropriate price for a unit of output. Once this price is determined, some means must be found to adjust the resulting revenues to customer class requirements. The traditional rate-making procedure begins with the amount of revenue necessary, then constructs rates to yield the required sum.

Average-cost pricing systems are claimed to be easier and less expensive to administer. Costs are easily determinable because they represent amounts paid or payable and not estimates of hypothetical future costs.

The uncertain relationship between revenue requirements and marginal costs has promoted the continuing use of accounting costs insofar as "there [has been] . . . no precedent, at least in North America, for basing the revenue requirement on marginal costs" (O.E.B., 1979, p. 10). The use of marginal costs was deemed to result in "substantially higher prices." In periods of inflation, marginal costs will lead to excessive revenues as long as revenue requirements are based on historical accounting costs. If marginal costs are in fact real costs and not abstractions, then both rates and revenue requirements founded on marginal costs should match production costs. The underlying fear is that those costs will be substantially larger than have been recognized in the past.

2.2.3 Summary and Conclusions

Many utility forecasts have been found to overestimate the growth in electricity demand. Canadian utilities at present have substantial excess capacity. The provinces, through their ownership or regulation of electric utilities, have been mainly responsible for overbuilding electricity supply (Helliwell, 1979). Many forecasts of energy consumption have incorporated high energy prices but not the potential effects of load management. Calculation of rates based on marginal costs appears to be inconsistent with traditionally determined revenue requirements.

2.3 HYDROELECTRIC SYSTEMS

The benefits of various types of rate reforms on hydro-constrained utilities depend on the amount of hydro power the utility has. In a report prepared for the U.S. Department of Energy, designed to simulate the effects of alternative rate designs, a "synthetic" hydroelectric utility was modeled based on composite data taken from the two major Oregon utilities that are predominantly hydro powered. It was found that "the more hydro power the utility has, the greater the benefits to rate forms which reduce overall electricity consumption and the lower the benefits of rate forms which shift load from one part of the day to another but do not reduce overall consumption" (U.S. DOE, 1979d, p. I-8).

The obvious reason for this finding is that hydroelectric systems are able to "store" electricity and therefore exhibit constant marginal costs

TABLE 2.2.   Relative Marginal Costs—Hydro-Constrained Utilities

	Spring/Fall		Summer		Winter	
	High Hydro	Low Hydro	High Hydro	Low Hydro	High Hydro	Low Hydro
Peak	$3.08	$3.19	$3.07	$1.66	$3.21	$8.28
Base	$1.29	$1.20	$1.00	$1.00	$1.74	$2.48

Source: U.S. DOE, 1979d, p. II-35.

over the day. On a seasonal basis, however, marginal costs could differ materially, depending on the water supply. During the dry season (i.e., the peak season), the marginal cost of supplying an additional kilowatt-hour will include the cost of marginally expanding storage capacity. All Canadian hydro-based utilities have their system peak in winter, when stream flows are at minimum. Further, distribution costs tend to be higher in hydro systems than in thermal systems.

A "hydro-constrained" utility is one that is unable to expand its power supply significantly because of the limited amount of hydro capacity available. In a hydro-constrained utility, marginal costs will exceed embedded costs by a much greater amount than that in thermal-based utilities. This is due to the low embedded costs of hydro capacity and because the marginal generating units are new, fuel-intensive thermal plants. As mentioned earlier, the greater the pondage capacity the utility possesses, the less marginal costs will vary by time of day.

Calculations of marginal costs for two synthetic hydro utilities are set forth in Table 2.2. The "high-hydro" case consists of 69% initial hydro capacity and the "low-hydro" case is 28% initial capacity. Note that in the season of system peak (in this case, winter), the time-of-day variations in marginal cost are less in the high-hydro case than in the low-hydro case.

There are two types of hydro power. Run-of-the-river hydro is generation by water-driven turbines without water storage. Pond hydro is generation using water stored behind a dam. The chief difference between the two is that pond hydro allows the utility more control over the time profile of electricity generation, whereas run-of-the-river hydro is much more subject to current availability of water. In both systems, if load shifting reduces total kilowatt-hour requirements, baseload capacity additions would be reduced and the cost savings would be the reduction in baseload capacity plus the variable cost of the generation eliminated. With run-of-the-river hydro, the benefits would be much greater for a load-leveling strategy.

The findings of the U.S. Foreign Rate Survey showed that the tariffs offered by hydro utilities were not different from those of conventional utilities (U.S. DOE, 1979c, p. VII-1). "Until 1951, Ontario Hydro's bulk-

TABLE 2.3. Electricity Use in OECD Countries, 1976 (thousands kWh per capita)

	Population Mid-1976 (Millions)	Household Use	Industrial Use	All Uses	Electricity As % of Total Primary Energy
Canada	23.143	2.98	5.14	12.28	38%
U.S.	215.118	2.80	3.35	9.91	32%
U.K.	56.001	1.52	1.79	4.60	34%
Sweden	8.219	2.42	4.97	10.52	40%
Norway	4.027	4.84	10.03	18.62	59%
New Zealand	3.116	2.69	2.25	6.71	44%
Germany	61.513	1.17	2.61	5.11	31%
Japan	112.768	0.87	2.71	4.34	20%

Source: Helliwell, 1979, p. 184. Reproduced, with permission, from the *Annual Review of Energy* 1979, Volume 4. © 1979 by Annual Reviews, Inc.

power charges were based only on demand [noncoincident peak] . . . because the costs of generating power in the hydraulic system did not fluctuate materially with the amount of energy supplied" (Ontario Hydro, 1976f, p. 1).

Analysts with the Long-Term Energy Assessment Program (LEAP) of EMR Canada state that most viable hydroelectric resources will soon be harnessed. Water power supplied 70% of Canada's electric generating capacity in 1968; by 1978 less than 60% was hydro; and by the end of the century this will be reduced to 33%. They project that a third of the electricity generated will be nuclear-fueled, compared to 9% currently, and that the remaining third will come from coal-fired plants (EM&R, 1978a, 1980).

The role of coal in electricity generation will be influenced by its alternative uses. It may be that the future role of coal lies with steel production, with exports, and as a source of synthetic fuels and petrochemicals (Helliwell, 1979, p. 199). If this becomes the case, the share of electricity generated by nuclear power will likely increase.

On a worldwide basis, hydro-endowed countries use more than average amounts of electricity, especially for industrial use. Figures for comparison are presented in Table 2.3.

Summary and Conclusions

Rate reforms based on time-of-use pricing and/or marginal costing methodologies will affect hydroelectric utilities according to the amount of pondage they possess. In general, rates that decrease overall consumption rather than flatten load curves benefit pond hydro utilities, whereas peak-load management benefits run-of-the-river hydro utilities.

Seasonal rates are more appropriate than time-of-day rates for hydro utilities.

2.4   RATE-MAKING CRITERIA

Rate-making criteria most often employed are based on the work of James C. Bonbright, *Principles of Public Utility Rates.* The core criteria are as follows:

Simplicity, understanding, acceptability, feasibility.

Proper interpretation, clarity.

Effectiveness in yielding revenue requirements.

Revenue stability.

Rate stability and continuity.

Fairness in apportionment of total costs.

Avoidance of undue discrimination.

Discouragement of wasteful use; promotion of justified use.

Rates can also be evaluated on other criteria. Bonbright adds additional considerations, such as electricity costs, energy consumption (changes in kilowatt-hours as well as changes in the use of other fuels), capacity requirements, and efficient resource allocation. This last criterion means that resources are efficiently allocated when the price of a commodity equals or reflects its marginal cost of production.

In evaluating the major marginal cost-pricing methodologies in use, the California Energy Commission employed the following considerations: (1) basis in economic theory; (2) reflection of cost causation; (3) applicability to California utilities; (4) documentation and comprehensibility; (5) ease of application; (6) sensitivity to input values; (7) degree of precision; (8) ability to evolve; and (9) usefulness for rate design (CEC, 1979, p. 8). Missing from this list is any consideration of the rate's ability to meet revenue requirements. In contrast, the Ontario Hydro ECAP Study stated that "the primary purpose of any pricing system and rate structure is to recover total operating costs" (Ontario Hydro, 1976f, p. 1).

The evaluation criteria chosen by Ontario Hydro were the following: (1) there should be an equitable division of costs between current and future years; (2) generally accepted accounting principles should be applied; (3) there should be a constant treatment of costs; (4) benefits of the method should outweigh the costs; and (5) the method should be versatile (O.E.B., 1979).

Any type of rate chosen must be evaluated according to a set of criteria, the choice of which is determined by the rate maker. Implementation of any one of the marginal costing methods is largely a matter of choosing those criteria that favor the use of marginal costs over traditional methods. If one values continuity, stability, and effectiveness in yielding revenue requirements, any departure from current practices will be overridden. Ambiguity in the meaning of criteria might cause arbitrary interpretations, as would unstated assumptions. For example, the criterion of "public acceptability" employed by Ontario Hydro was expanded to mean consistency with other government energy policies (O.E.B., 1979, p. 13).

The rate-making criteria most evoked in debates on marginal cost pricing involve (1) the revenue and rate stability arguments and (2) issues of fairness, efficiency, and income redistribution.

2.4.1   Revenue Requirements and Rate Stability

Traditionally designed rates will always yield revenue requirements because they are designed to do so. The issues are whether another rate structure can meet revenue requirements and whether these alternative structures will prove to be more or less stable than those in effect.

Marginal cost rates are taken to be synonymous with time-of-use rates. The advantage, proponents claim, is that revenues from additional sales will, by definition, cover the cost of additional sales, and the reduction in revenues from reduced sales will equal the reduction in total costs. If rates do not automatically follow rising marginal costs, erratic price increases will result (Wisconsin PSC, 1979f, pp. 3–4).

The disadvantage of time-of-use rates is that they are extremely weather sensitive; revenues might fall during an unseasonably cool summer or warm winter. In Canada this problem is avoided because regulation usually involves a "test year" calculation that abstracts from any abnormal or random circumstances that cause fluctuations in revenues (Helliwell, 1978, p. 118). Another solution might be a stabilization fund, similar to that maintained by Ontario Hydro.

Rates based on peak-load pricing would also tend to shift as customers changed from peak to off-peak usage. The value to the utility of initial shifts off peak is very large. However, as load curves flatten, the subsequent savings diminish. During the course of the shift in the load curve, the peak/off-peak differential would have to be changed to reflect the new values of off-peak consumption.

2.4.2   Fairness, Efficiency, and Income Redistribution

In its most general sense, fairness means that a customer should pay what the service costs. This can be distinguished from equity, which means that rates should not be discriminatory either between customers within a class or between classes of customers. Time-of-use pricing is considered fair inasmuch as those most responsible for creating peak system loads absorb the costs they impose on the system. This raises the issue touched on earlier as to the precision of responsibility for system peak. The ECAP Study "stated that fairness meant there should be no seniority rights in pricing electric service, . . . [as] all consumption is always new" (O.E.B., 1979, p. 12). One report has suggested that fairness should not be taken too far as a rate-making goal. Efficiency was to be valued over fairness because if a tariff system proved to be unfair, this could be remedied by other transfers without impairing efficiency. If, however, efficiency is lost through the rate structure, it would be harder to remedy (Ontario Hydro, 1976e, p. 4).

Questions of efficiency hinge on the proposition that efficiency is maximized when all goods are priced at their marginal value. One question is the extent to which efficiency would be furthered if it were adopted as a pricing objective. The issue became confused by two conceptions of efficiency—one economic and the other operational, that is, engineering efficiency representing the full utilization of plant and equipment. In Ontario it was decided to pursue engineering efficiency and reject economic efficiency: "economic efficiency cannot be achieved through electricity rate structures. . . . [it] is not a valid pricing objective" (O.E.B., 1979, p. 27). Absent marginal cost pricing of substitute fuels, no case was made that economic efficiency could be furthered by repricing electricity alone.

Associated with the fairness criterion was the implication that the distributional effects of any rate structure should be economically neutral. Economists involved with rate reform have consistently rejected the notion that utility rates are a proper channel for redistribution of wealth. "Income redistribution in particular is not an appropriate rate design goal but a function of government . . ." (O.E.B., 1979, p. 59). If surplus revenues do result from imposition of marginal cost pricing, the implication is that the surplus should be redistributed to the public in a manner that least disturbs the current distribution of wealth.

These arguments will surface again when lifeline rates are discussed later in this book. The poor, who are most affected by increasing energy costs, should be protected from any potentially adverse effects of rate reform. Despite the fact, however, that rate-making criteria are sensitive to questions of equity, the object of rate reform is the achievement of a better correspondence between cost and price. The more functions a given rate is asked to perform, the less likely it is to be cost-based.

2.4.3 Summary and Conclusions

Rate structures are evaluated by a set of criteria chosen according to rate makers' preferences. The choice of criteria will determine the acceptability of a given structure. It is uncertain whether rates based on marginal cost would be more stable than rates currently in use because the effects of load shifting would be added to the disruptive influence of rising fuel prices. The concepts of both fairness and efficiency relate to service charges that reflect cost; income redistribution is rarely cited as an appropriate goal of rate making.

3

ELECTRICITY PRICING:
RATE SETTING

3.1 THE TRANSITION FROM COSTS TO RATES

3.1.1 The Allocation of Costs to Periods

The joint costs incurred in providing electric service at different times, which must be allocated to particular units of service, is a perplexing costing problem. The three issues raised in connection with the allocation of utility costs to time involve the use of loss-of-load probability, the duration of the peak period, and the recurring question of user responsibility for peak.

Loss-of-load probability (LOLP) is the most commonly cited method for the allocation of demand-related generation costs to rating periods. It is based on the probability of outage in each hour. The costing method, developed by National Economic Research Associates, Inc. (NERA), makes use of generation plans and operating practices of a utility as well as specifications of future capacity additions and maintenance plans (Ontario Hydro, 1976g). The value of loss of load at any given hour in this formulation is based on the costs of extending the system to meet the shortage.

In an alternative use of LOLP, the shortage cost of outage is calculated. Theoretically, the marginal cost curve can be envisioned as increasing with running costs until capacity is exceeded; the cost curve then becomes the shortage cost to customers losing power. A variation of this method is employed in France. The French national utility calculates loss to customers from a hypothetical shortage. It then plans capacity additions until the cost of the last unit of capacity added equals the probable cost of a failure (Ontario Hydro, 1976g, p. 4). By North American standards this is a highly unorthodox approach.

A constant loss-of-load probability is a standard of reliability, but the measure itself does not account for differences in the duration or extent of a possible outage. These differences are being explored through research, and estimation of outage costs as a viable measure may one day become acceptable practice (Cicchetti and Reinbergs, 1979, p. 240). It should be evident that loss-of-load probability is most sensitive to changes in peak load. In a model based on data from an oil-fired generation plant and a coal-fired one, reducing the peak load by 8% reduces the loss-of-load probability by approximately 80% (U.S. FEA, 1976b, p. 37). This should imply that a utility experiencing growth in its baseload with peaks reduced would require fewer capacity purchases (either as purchased power or additional capacity).

In power generation, each increment of demand on the system implies a deterioration in reliability, so it is the increment in demand that is important rather than the overall level of demand. Use of marginal loss-of-load probability implies that the contribution to a probability of disrupted service increases as peak capacity is approached.

If one considers the responsibility for increasing loss-of-load probability as a graduated peak responsibility, the relative value of the LOLP reflects hour-by-hour increases in demand rather than heightened costs for the peak hours. NERA argues that every peak user actually imposes on society, in the long run, the cost of incremental capacity. There is no such causal connection between off-peak use and capacity costs, they argue, because the capacity would exist whether they used it or not (Ontario Hydro, 1976e, p. 9). Following this line of reasoning leads one to conclude that peak users should be responsible for *all* long-run additions to capacity necessitated by their demands, whereas off-peak users should be responsible for no incremental capacity whatever. NERA has suggested the use of a modified LOLP measure, "probability of contribution to peak," which advocates that demand costs be allocated to peak rating periods and none to off-peak periods. The effect of this proposition would be to create wide price differentials between peak and off-peak rates.*

A counter argument is that systems are expanded not only to meet increasing peak demands but also to meet growth in off-peak demand

---

*Calculations performed for Ontario Hydro show that nighttime LOLPs are about 1/10 to 1/100 that of daytime LOLPs (Ontario Hydro, 1976g, p. 9).

while maintaining a specified reserve margin. Therefore, a reduction in overall consumption, not just peak period demand, would result in net savings to the utility from delaying a system expansion plan. Under these assumptions, the peak/off-peak differential would be much narrower, that is, peak users would not be held liable for all capacity additions.

The loss-of-load probability data introduced at Ontario Hydro hearings indicated an extreme volatility in load trends resulting from the greatly revised load forecast (i.e., system overcapacity) and differences in maintenance scheduling. This volatility, it was concluded, "could seriously affect the allocation of demand costs and ultimately rate stability if LOLP data alone are used" (O.E.B., 1979, p. 40).

A second problem in the selection of rating periods is the determination of what constitutes the peak period. Because all rates are in some measure weather determined but must be selected in advance, the period chosen can only reflect the probability of weather conditions. This choice may or may not respond to actual demand conditions.

The next problem is the length of the rating period. If the peak is narrowly defined (e.g., the number of hours is within 5% of the expected peak), there will be a high cost per unit of demand associated with those few peak hours. Conversely, where there is a broad peak, the peak/off-peak price differentials might have little impact. A narrowly defined peak tends to cause demand to be shifted off-peak, creating a new peak at the edge of the first. This phenomenon has been observed in Europe (Mitchell and Acton, 1977, p. 47; and Acton et al., 1978c, p. 261).

In practice, marginal costs are averaged over the hours constituting the peak period. Carpenter and Osler have pointed out that the 16-hour daily peak period proposed by the Ontario ECAP Study was too long and would result in too much cost averaging. This consideration was to be balanced against more and shorter rating periods, which would reduce the amount of cost averaging yet complicate billing and confuse customers (O.E.B., 1979, pp. 38–39).

A final point to be decided by rate makers is the duration of the rate schedule itself. As discussed earlier, this concerns the issue of whether short- or long-run marginal costs are used as a basis for calculations and how rapidly system costs are expected to change in response to pricing incentives.

3.1.2   Allocation of Costs to Customer Classes

A related issue to the selection of time periods in the design of rate schedules is whether to maintain the convention of customer classes under a reformed tariff schedule.

When non-time-differentiated accounting costs are used, costs are allocated to customer classes as broad consumption categories. The costs that an individual member of a class imposes on the system may be different from those attributed to his class. Using time-differentiated rates, marginal costs for each rating period are calculated by multiplying the costs for each period by the unit of electricity sold to each customer class. The revenue requirements are the sum of the costs for each class, and the rates are constructed to meet the revenue requirements.

It is now being argued that customer classes are an artificial classification, that is, costs for energy and demand should be uniformly applied across customer classes. If rate schedules are to be based on daily load patterns, a residential and commercial customer with the same pattern of demand should be subject to the same rates. This would ensure that rates were fairly applied across classes as well as between individual members of the same class. The marginal cost tariffs in effect in Europe have eliminated the provisions of earlier tariff structures that discriminated among customers according to the uses to which they put electricity, or whether they were commercial or industrial. The charges in effect are based directly on the electrical usage reflecting system costs. A variety of tariffs is made available to all customers supplied at the same voltage, and the customer is free to pick the rate that best suits his use from the available choices (Mitchell and Acton, 1977, p. 4).

It should be recognized that load characteristics differ by customer class. Figure 3.1 illustrates that industrial loads are relatively flat compared to those of residential and commercial customers. Although commercial firms appear to cause the most variation in system load, they were found to be the least able to adapt their usage to off-peak rates. These establishments, which include hospitals, schools, hotels, stores, etc., normally operate during business (on-peak) hours (Ontario Hydro, 1976j, p. 6).

3.1.3 Summary and Conclusions

Loss-of-load probability is a generally accepted method for relating customers' demands on a power system to the costs of meeting demand on an hourly basis. It can be argued that peak-period users are responsible for all costs of system expansion or that peak and off-peak users share responsibility for baseload capacity and peak users for peak capacity only. The peak/off-peak price differential is influenced by allocation of underlying capacity costs, as well as the length of time that peak rates are in effect. Narrowly defined periods cause "peak chasing," the creation of a new peak, while broadly defined periods diminish the effects of peak/off-peak price differentials. Tariff schedules should be amended to re-

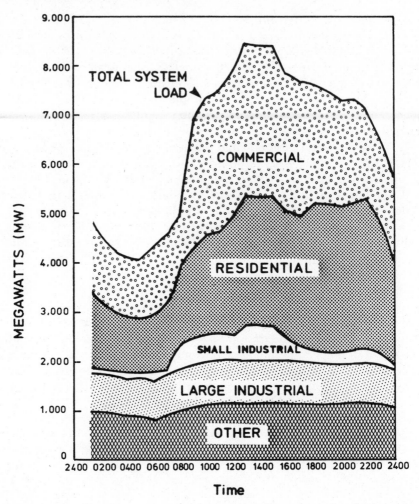

FIGURE 3.1. Sample load curve showing consumption by customer class for the summer season in a peak weekday. (Source: U.S. FEA, 1977a, p. 30.)

move the distinction between customer classes and to reflect only the quantity and time of power demanded at similar voltages.

3.2 MARGINAL COSTING METHODOLOGIES

The most widely espoused costing methodologies are reviewed in this section. Many of these were tested by the California Marginal Cost Pricing Project, and all were ultimately rejected. The two principal grounds for rejection were (1) that none had a well-developed definition of the load increment to be evaluated and (2) that none reflected the planning process that links demand and cost (CEC, 1979, p. 90).

The California Marginal Cost Pricing Project, funded by the U.S. Department of Energy, tested each of the methods on data provided by California's three largest utilities and then derived hypothetical rates based on each method. Unfortunately, the rates are not directly comparable because the assumptions underlying each of the approaches differ; for example, some have three-part rates with customer charges and energy and demand charges, whereas others have only demand charges or only energy and demand charges. No attempt was made to compute an "average" bill for a particular class of user or the overall effects on utility revenue requirements. The exercise was directed at translating marginal costing theory into practice and evaluating the resulting rates against the criteria outlined above.*

3.2.1   The NERA Method

The NERA (National Economic Research Associates) method defines short-run marginal cost as the cost of energy for the time served plus the shortage cost for the time not served. Long-run marginal cost is defined as the energy cost plus the cost of capacity at peak. Optimally, short-run capacity costs should equal long-run capacity costs. The NERA method is based on viewing the system as optimally adjusted to meet increased peak demands at least cost. The marginal unit is therefore assumed to be the least-cost unit, and marginal generating capacity costs are defined as the annuitized cost of the plant used to meet the load (usually the peaker).

NERA uses a utility's own accounting-type cost categories and the method has not yet been computerized, although it can be. The data collection task is formidable and detailed.

Marginal energy cost at any time is composed of the fuel and variable operating and maintenance (O&M) costs of the last unit on line, which normally is the generator with the highest running cost. This cost concept is called the system lambda (λ). System lambda is derived from the following mathematical formulation. Total operating costs are minimized subject to the constraint of meeting a particular load that might occur:

---

*The descriptions of methods presented here are of a cursory nature only. In fairness the original sources should be consulted. These references are listed in the bibliography. Much of this chapter relies on the critical evaluations of the California Marginal Cost Pricing Project.

Minimize $F(EC) + \lambda(L_i - Q)$

where $F(EC)$ total running cost
       EC energy cost
       Q output
       L_i load at time i
       λ LaGrangian multiplier

Then $\partial F(EC)/\partial Q - \lambda = 0$. Since $\partial F(EC)/\partial Q$ is the change in total running cost with respect to a change in kilowatt-hour demand, λ is the marginal energy cost per kWh (CEC, 1979, p. 18).

The NERA method adjusts for capacity-related losses (which occur after the system is adjusted to meet a new level) and demand-related O&M expenses.

These marginal running costs (fuel and O&M) are usually taken from the utility's production cost modeling program, reduced to an hourly basis and grouped into periods corresponding to the predefined pricing periods. The hourly marginal costs are weighted by the total hourly load to derive a weighted average marginal energy cost for each period. The central tendency of these costs over a five- to ten-year period becomes the marginal energy cost for each price period, so the marginal energy cost used for rate making is actually an average of marginal energy costs over the planning horizon.

Transmission and distribution costs are also calculated on a marginal basis, including transmission capacity cost (dollars per kilowatt of system peak resulting from expanding the transmission system in the most efficient manner possible), and marginal distribution costs (broken down into customer-related connections, distribution investment above minimum demand system, operating and maintenance costs, and energy-related distribution costs calculated as energy losses). Other costs calculated include administrative and general expenses, plant loading factor (expanded office facilities and transportation equipment), working capital expenses, and a carrying charge adjusted for inflation and technical progress (unique to the NERA method).

The allocation of costs to costing periods is done via loss-of-load probability (LOLP). This is based on the economic logic that customers will pay no more for electricity than the cost of foregoing electricity, that is, the cost of a shortage. The expected yearly marginal shortage cost is defined as the sum of the hourly shortage probabilities multiplied by the periodic marginal shortage costs:

$$C = \sum_{i=1}^{n} p_i \, S_i$$

where C = marginal capacity cost per year
 $\quad P_i$ = probability of shortage in period i
 $\quad S_i$ = shortage cost in period i
 $\quad n$ = number of periods

If the shortage cost is assumed constant, each period's relative contribution to annual capacity cost is proportional to its probability of shortage. This concept is frequently encountered in other costing methodologies as well. Capacity costs are therefore allocated via loss-of-load probabilities as follows:

$$F_i = \frac{P_i}{A}$$

where F_i = cost allocation factor for period i
 $\quad P_i$ = LOLP in costing period i
 $\quad A$ = annual LOLP

The use of LOLP by NERA and others is a controversial issue. If it is argued that capacity is added because of the utility's potential inability to meet an increase in load, that inability to meet load (the LOLP) should measure the need to add additional capacity. The cost of that additional capacity could then be considered as marginal capacity costs. Opponents of the concept argue that actual LOLP is very small, generally a few hours per year (or almost nonexistent with high volumes of excess capacity), and its use is not an appropriate basis for time-of-use rates.

The Ontario Electricity Costing and Pricing Study (ECAPS) initially adopted a method developed by NERA that assumed that, in the case of Ontario Hydro, the marginal demand cost would be the capital cost of a combustion turbine unit (the least capital-intensive unit that could be added to the system), assuming optimality. When the nonoptimality of the system was apparent, other methods were advocated for estimating marginal demand costs. Among the methods suggested were (1) using marginal shortage costs (i.e., the opportunity cost of a kilowatt); (2) price experimentation to determine at what point demand would be rationed to available supply; (3) the use of econometric studies or elasticity estimates to predict the price of capacity in the export market; (4) the study of the value added of electricity; and (5) the evaluation of the losses customers would suffer in the event of an outage (O.E.B., 1979, p. 47). Partisans of these alternatives realized that their methods did not calculate marginal cost in the theoretical sense, but rather, incorporated compromises to reflect the real situation. The variety of methods attests to the economists' resourcefulness in finding a solution. The problem rests, however, with NERA's assumption of optimality, which is more often than not violated in real-world conditions. However, electric utilities in

France and Sweden base their marginal cost calculations on the theoretical system that would be optimal (i.e., constructed to operate at minimum cost), ignoring the suboptimality of the current mix (Mitchell and Acton, 1977, p. 10).

The ECAP study, using the NERA methodology, estimated peak energy costs as a weighted average of energy costs of a combustion turbine unit and imputed peaking hydro costs. "This approach was criticized on the ground that the actual marginal running cost of peaking hydraulic generation is close to zero" (O.E.B., 1979, p. 46). This relates to the notion that peak period charges should include the costs of all underlying baseload capacity, a position favored by NERA.

3.2.2 The Cicchetti, Gillen, Smolensky Method

Cicchetti, Gillen, and Smolensky (CGS) calculate marginal generation capacity costs on the assumption that the marginal unit of generation capacity is the next plant in a utility's expansion plan (based on 10–20-year forecast) that would be built or not, accelerated or delayed, in response to a change in the expected demand level. The annuitized value of the marginal plant divided by the size of the plant is taken to be the marginal cost of a kilowatt of generation capacity. Thus, marginal generating cost is that resource plan modification that the utility chooses to meet incremental demand, that is, the change between possible alternative scenarios for the future. If additions to the expansion plan involve additional gas turbine peaking units, marginal generating capacity cost is the annuitized cost of the unit itself. These units do not produce fuel savings (as in the case of larger, more efficient intermediate and baseload units) (Cicchetti et al., 1977, p. 9). For a given hour, the marginal energy cost is the cost of the least efficient generator on line. Capacity costs are allocated to time periods using LOLPs.

The marginal cost of generation can assume a number of forms. If the utility purchases additional power, the marginal cost is the price paid for the new capacity. If the decision involves accelerating or delaying a plant, the cost becomes a stream of costs or benefits extending into the future. If a plant is deferred by one year, this implies a permanent savings.

The marginal costs of energy at any hour are the fuel and variable operating expenses of the last unit on line—the system lambda. Transmission and distribution costs are calculated by estimating the effect on system costs of a change in kilowatts demanded at each voltage level. Customer costs are not included in the CGS calculations but are left for each utility to calculate on an ad hoc basis. Cicchetti et al. suggest that such calculations be based on replacement costs for the equipment used to serve a given class of customer.

To allocate costs to time periods, demand in relation to available capacity is estimated using LOLP. The narrower the margin between demand and available capacity, the greater the probability that load will exceed capacity or that demand in that period will require additional capacity. Theoretically, the hourly probabilities will cluster into seasonal and daily costing periods.

Criticizing the CGS method on economic grounds, the California Energy Commission found that basing calculations on a resource plan and additions to capacity makes the identification of a single plant as the marginal plant difficult and perhaps arbitrary, nor is the solution obtained economically efficient. In order for economic efficiency to prevail, the current resource mix of the utility must be optimal and the projected expansion must also move the utility toward an optimal mix. If the current plant is not optimal, the projected additions will be different from those chosen under optimal circumstances. The NERA method assumes optimality and thus avoids this problem. The evaluation by the California Energy Commission did find that the estimate of marginal costs derived from the CGS methodology is helpful to the consumer, however, because it does reflect the changes in costs over time (CEC, 1979, p. 46).

In practice, it is quite difficult for a utility to alter its pattern of planned construction. "It must be realized that a public utility with a sizeable construction program does not have the flexibility to postpone or advance the program at will" (*The Financial Post*, 1980, p. 29). British practice, however, uses incremental costs calculated by hypothetically accelerating the date at which the next generating plant will be added. This is essentially the CGS method (Mitchell and Acton, 1977, p. 10).

3.2.3 Gordian Associates' Method (Optimization)

The Gordian Associates' approach employs a mathematical model that minimizes costs subject to system reliability and financial constraints. It is essentially a computer simulation of utilities' response to load changes.

Short-run marginal costs in this model are defined as the avoidable costs incurred by a utility that are caused by customers' marginal consumption. The costs are a combination of energy cost and marginal operation and maintenance costs. A marginal demand charge is calculated on the basis of a price necessary to limit customers' demand to a level that maintains system reliability.

Marginal generation costs are calculated via an optimization routine that minimizes the present value of all future utility system costs subject to reliability constraints. The program is further limited by necessary financial constraints and revenue requirements. An added benefit of the system is that pro forma financial statements are produced. The entire Gordian program is computerized, although the program belongs to

Gordian Associates and empirical tests have not been made public. Supplemental subroutines in the model determine least-cost operating schedules (dispatch), system reliability, and maintenance schedules. Loss-of-load probability is used as the index of reliability. Estimated reserve margin and LOLP are used as inputs to the main generation expansion model (an optimizing linear programming model), which minimizes the present value of utility system costs while calculating annual capacity additions.

The output of the LP model provides the following: (1) present value of all utility generation system costs that correspond to the optimal capital addition plan; (2) generation capacity additions by type of plant and year; plus (3) data on the percent utilization of each plant, the capacity of storage units, and interchanges with other utilities of capacity and storage.

Marginal capacity is defined as the "optimal way of providing additional capacity" in each year, so the program might specify additional nuclear capacity in a given year as the most efficient and hence "marginal" capacity addition (CEC, 1979, p. 76). Transmission and distribution costs are calculated separately and the model has no explicit method for estimating marginal customer costs, which, Gordian argues, are not properly considered as marginal costs.

The marginal generation demand costs that result from the optimization model are allocated to time periods using loss-of-load probabilities. The LOLP, as usual, is used to create a weighting index to indicate capacity required to maintain reliability. Costs by time periods are obtained by multiplying the gross cost per kilowatt of capacity by the economic carrying charge in order to estimate net marginal cost per kW-year.* The net marginal cost is then multiplied by the LOLP weighting index to get a marginal cost per rating period.

The model has been criticized because its assumptions and limitations are not well known due to the fact that the programs have not been publicly available. The use of LOLP, common to most methodologies, remains a problem when costs must be allocated to time periods (CEC, 1979, p. 83).

3.2.4   The EBASCO Method

The EBASCO Marginal Cost Methodology, or the base-intermediate-peak method, is not properly a marginal cost method, but rather a means

---

*Standard carrying charges translate the present value of total fixed costs that will be incurred into an annual fixed amount that is paid by rate payers over the life of the plant; the Gordian economic carrying charge includes special factors to adjust the standard carrying charge for cost increases over time, technological improvements, changes in the relative prices of fuels, and changes in demand patterns.

of estimating average future costs and allocating them to the three periods—base, intermediate, and peak (B-I-P). Long-run marginal cost is
defined as the average cost of serving new requirements in the long run.
The capacity costs are the present dollar cost per kilowatt of the total
planned system additions over the planning horizon. Energy costs are
the average expected costs into the future using both existing and
planned capacity. Fixed annual costs are levelized over five years, rather
than the life of the plant, and then allocated to the rating periods using
the B-I-P method.

The three costing periods each have different costs because of the
type of generation used to meet demand in each period. The costing
periods can be derived by employing a load duration curve that indicates
the type of capacity used to meet demand in each period.

The method is noncomputerized and may be performed by the consultant with utility personnel assistance. Because the method is chiefly
concerned with reflecting proportional cost causation and uses projected
average energy and supply additions rather than marginal costs, it is
more a mechanism for peak-load pricing than marginal costing. It seeks
to match the costs of a period to users' responsibility for causing the
costs. Base capacity is charged to all time periods. Total annual capacity
costs are assigned to base, intermediate, and peak periods, and each new
generating unit is identified as belonging to one of the categories. If the
costs fall into two rating periods (peak and off-peak), the cost by generator type would be classed as follows:

Capacity Cost by Generator Type	Cost Assignment to Rating Period	
	Peak	Off-Peak
Peak	All	None
Intermediate	$\frac{1}{2}$	$\frac{1}{2}$
Base	$\frac{1}{2}$	$\frac{1}{2}$

Although this cost allocation procedure has been criticized as being
simplistic and arbitrary, only peak costs are assigned to the peak period
and peak-period users are accountable for only the direct costs they
impose on the system, rather than all capacity costs (CEC, 1979, pp. 87–
88).

The California Marginal Cost Pricing Project task force found the
EBASCO methodology the least appealing of all the marginal cost approaches and consequently did not test it empirically. The methodology
was rejected primarily on the ground that it was an average-cost method
of deriving time-of-use prices, and not a marginal cost method per se.
Nevertheless, it was found to be widely applicable, understandable, and
relatively simple to use. It has been judged "very good for load factor

improvement but poorer for revenue sensitivity and rate stability" (O.E.B., 1979, p. 99).

A modified version of the base-intermediate-peak (B-I-P) method has been proposed. The original B-I-P method designed by EBASCO apportioned baseload costs equally to base, intermediate, and peak periods, reasoning that these facilities supply the same demand requirements to each rating period. The modified B-I-P method allocates all capacity costs on the basis of the energy produced by each type of generation in each costing period. The cost allocation procedure includes the number of hours in each period, whereas the original EBASCO method allocated costs on predetermined percentages without regard for the number of hours in each period (O.E.B., 1979, p. 94). This modified version can be visualized by reference to a screening curve that plots the relative costs of each type of capacity against time. The resulting curve (Figure 1.4) is a least total cost combination by hours of use.

3.2.5 The California Marginal Cost Pricing Project (MCPP Method)

In rejecting each of the previously described methods as a means of marginal cost pricing, the California Marginal Cost Pricing Project set about to create its own method. To devise a method that would be free of the shortcomings of the other approaches, the Pricing Project proposed a "scenario" method. This method would determine which costs change in response to a given *load* scenario. The seven scenarios that would be evaluated are load increases in (1) all hours; (2) all summer hours; (3) all winter hours; (4) summer on-peak hours; (5) summer off-peak hours; (6) winter on-peak hours; and (7) winter off-peak hours. The utilities would then be forced to identify their responses to changes in these anticipated future loads so that an actual change in utility costs could be derived by *time of use*. The costs obtained would relate to different service periods corresponding to the seven scenarios (CEC, 1979, pp. 91–92).

This method tends to skirt the issue of using LOLP as an external cost allocation method. The LOLP method of deriving rating periods can be "subjective and not necessarily related to costs or loads" (O.E.B., 1979, p. 90).

The scenario method is defended by the MCPP as follows:

> The definition of marginal cost inherent in the scenario approach is the change in total costs that will result from a change in demand. These costs are calculated using the utility's actual operating costs and resource plan, and therefore appropriately capture the marginal cost pricing concept plus a rationing charge to maintain efficient long run system reliability (CEC, 1979, p. 93).

The method relies on the utilities' resource planning tools and practices, which cannot be incorporated into a uniform model applicable to different utilities.

3.2.6 Summary and Conclusions

Eventually, a marginal cost methodology will be developed for electric utilities that will have broad applicability. Such a method must be readily understandable to all participants. It should be formulated so that data requirements are uniform across utilities and data are neither onerous to obtain nor unreliable. The program must be flexible enough to incorporate changes in future capacity plans and changes in input prices, as well as to incorporate the responses of customers to time-of-use rate schedules. Such a program could serve to supplement forecasting efforts, evaluate alternative energy supply and conservation options, and perhaps even lead toward consistent electricity pricing among regions.

3.3 ADJUSTING RATES TO REVENUE REQUIREMENTS

Marginal cost pricing will lead to excess revenue in times of rising prices as long as revenue requirements are based on historical accounting costs. The problem of distributing these excess revenues is another difficult question in formulating a marginal costing system. To meet and not exceed current revenue requirements, these funds can be returned, credited against future consumption, or kept to finance some public program.

To the extent that rates are adjusted to reflect current revenue requirements, they will depart from marginal costs and, in practice, defeat the purpose for which they were theoretically established. Further, the greater the departure from cost, the more the rate-making process becomes politicized and akin to taxation.

A number of methods of surplus redistribution are reviewed briefly. Oddly, it has not been suggested that rates based on marginal costs would ever fail to satisfy current revenue requirements, so a surplus is always assumed. Whether we will ever return to a time of decreasing generation costs with volume (i.e., increasing returns to scale) is doubtful. Cicchetti et al. (1977, p. 15) state that "the real cost of generating capacity (i.e., the cost net of purely inflationary effects) appears to be rising."

3.3.1 The Inverse Elasticity Rule

Using the inverse elasticity rule to adjust marginal costs to revenue requirements involves setting different prices in different markets so that the "percentage deviation of price from marginal cost in each market is inversely proportional to the elasticity of demand in that market." For example, price "would be lowest in the least price-elastic market and highest (closest to long-run marginal costs) in the most price-elastic market" (Ontario Hydro, 1976e, p. 14). Using the inverse elasticity rule, the initial block ("lifeline" block) is presumed to be less price elastic than usage in the residential tail blocks or commercial and industrial usage. As a result, reducing the price to low-use customers and raising the price to high-use customers will minimize deviations from the appropriate levels of usage. The reasoning is as follows: In markets where demand is most elastic (most responsive to price changes), a price set nearest to marginal costs will discourage excess consumption in circumstances where all rates have been set at less than marginal costs in order to avoid excess revenues. Because we have been assuming that rates set at less than marginal cost do not cover the value of resources used in electricity generation, demand should be discouraged where users are most able to adjust their consumption. Where usage is least responsive to changes in price, excess revenues are allocated where consumption will not, by definition, be much affected.

In standard price discrimination, which a textbook monopolist would employ to exploit the total area under the demand curve, higher prices would be charged for the most inelastic demand and correspondingly lower prices would be charged for the more elastic demand. The inverse elasticity principle is the opposite of this. If necessary, however, it could be reformulated to generate additional revenue by approaching the monopoly model. The inverse elasticity rule is used in France when a deficit is projected. Customers with a low ratio of average to peak demand (i.e., low load factor) are charged the greatest increases (Mitchell and Acton, 1977, p. 4).

The inverse elasticity rule has been criticized on a number of grounds. First, it represents a form of price discrimination against high-volume users, who can be expected to oppose its imposition vigorously. Second, it requires knowledge of the elasticity of demand for each customer class, data that are difficult to acquire and often unreliable. It is also common to find that differences between elasticities of demand within customer classes are as great as those between different classes of customers.

3.3.2 Uniform Dividend

Under this method the amount of surplus is evenly distributed to each customer like a dividend, regardless of level of usage. The weaknesses are apparent. First, a small user would reap relatively greater benefits than a larger one. In this sense it has income redistribution consequences. Second, the reduction in the bill might serve as an incentive to increase energy consumption.

3.3.3 Equal Percentage Reduction

The equal percentage reduction plan would apply a proportional rebate relative to the customer's average bill. If the rebate were 10%, a monthly bill of $50 would be refunded $5, and a monthly bill of $50,000 would be refunded $5000. Because deviations from marginal cost are uniformly applied, the method is nondiscriminatory, but the method tends to bring rates back to a level that approximates average cost.

3.3.4 The Benchmark Rate

The benchmark rate is more properly a rate structure in itself than a means of rate adjustment, but it does tie rates firmly to revenue requirements. The method, proposed by Cicchetti, is designed so that any increase or decrease in the use of energy over the initial benchmark period (usually three years) is priced at a given rate. This same rate is also used to add or deduct charges for increased or decreased usage. Use during the benchmark period is multiplied by a flat rate designed to match the revenue requirement for the customer class (Cicchetti and Reinbergs, 1979, p. 252).

This method enables customers to see the full marginal cost of additional consumption by reflecting these costs in the latter blocks and determining the length of the blocks by prior usage. Revenue requirements would be met and increased consumption priced at or near its marginal cost.

Other methods for dealing with the surplus involve imposing a lump-sum tax on the utility, equivalent to the surplus. An alternative proposed by Helliwell would use the surplus to help finance the investment of energy conservation devices. Alternatively, the payment of a rebate could be made conditional on the user meeting certain minimum insulation or other efficiency standards (Helliwell, 1978, p. 126).

3.3.5 Summary and Conclusions

Under the assumptions that marginal costs will lie above average (accounting) costs, a surplus in revenues will result from using marginal cost-based rates. The objective is to return this surplus to consumers so as not to increase their energy consumption or cause utility bills to deviate from the original principles of marginal cost. Application of the inverse elasticity rule lowers rates in the initial blocks, where demand is least elastic. Other methods return the surplus to consumers via equal lump sums or according to prior periods of consumption.

Ultimately, however, some equilibrium should be reached where rates based on marginal cost equal true costs of service and an excess is no longer generated. Before this stage can be achieved, cost accounting itself must be marginal cost based rather than historical cost based.

3.4   ALTERNATIVE RATE STRUCTURES

3.4.1   Declining-Block Rates

Declining-block rates have tailblocks that are below average costs. Historically, this structure reflected economies of scale. The rate structure actually promoted increased consumption so that all customers' costs could be reduced. Declining-block rates are characterized by a high initial block, where consumption is least sensitive to price, which ensures recovery of fixed expenses.

Declining-block rates are not designed actually to track costs, nor do declining costs necessarily justify decreasing rate levels with increased consumption. "The fact that marginal costs may be below average does not in itself justify confronting high-volume customers with a lower marginal charge than low-volume customers. . . ." (Ontario Hydro, 1976e, p. 14). Residential customers, under declining-block rates, shouldered the burden of historical investment through the initial demand charge.

Demand charges do little to discourage consumption, nor do most consumers know what constitutes the basis for their demand charge. The declining tailblock, however, does encourage consumption, as this is the more elastic part of the schedule.

Conceivably, there could be a cost-based declining-block rate if the tailblock only reflected fuel costs and no capacity costs. For example, the widespread adoption of fuel adjustment clauses has brought tailblock rates more in line with costs so that increased sales will not magnify utility losses.

An argument used to justify declining-block rates is that customers with higher load factors are less expensive to serve because less plant and equipment are required. The marginal cost of serving additional industrial customers is well below that for residential and commercial customers (Christensen and Green, 1978). Industrial customers, as we have seen, also impose less variation on the utility's load curve than other customer classes. If large customers within the residential class also have higher load factors, this would constitute a justification for declining-block rates. If, however, low-usage customers had higher load factors, there would be a cost-based justification for inverted rates. In fact, although load factors change with usage, there is no proven direct relationship between the two (U.S. DOE, 1979d, II-14).

Declining-block rates are also an efficient method for capturing consumer surplus (i.e., the area under the demand curve above price). Rates are lowered along the elastic portion of the demand curve to utilize capacity fully. This in turn may promote more capacity expansion.

3.4.2   Automatic Fuel Adjustment Clauses

Automatic fuel adjustment clauses are a method of maintaining utility revenue stability. Because unplanned increases in the cost of fuels cannot be easily financed, the adjustment clause has proved to be a convenient mechanism for meeting these expenses. Coal and nuclear fuels are less subject to fluctuations because they have been purchased under long-term, 10–40-year contracts, although increases in the prices of these fuels will also be felt as contracts expire. The use of fuel adjustment clauses has dampened utilities' interest in rate reform as increased costs can be passed on directly to consumers, often without time-consuming rate hearings. It is argued that the continued use of such clauses offers utilities no incentive to minimize their fuel costs, but the effect of the clauses on plant expansion is less clear.

Some argue that the use of fuel adjustment clauses will not lead the utilities to seek more fuel-efficient generating capacity because they are in effect sheltered from the effects of rising prices. Empirical research conducted by Atkinson and Halvorsen in 1976 studied the effects of automatic adjustment clauses (or "tracking") on the input decisions made by regulated utilities. "To introduce tracking for an input whose price is rising will increase the use of that input relative to others. . . . If fuel costs are tracked . . . utilities will increase their use of fuel relative to other inputs whose cost increases are not tracked" (Helliwell, 1978, p. 117). (See also Atkinson and Halvorsen, 1980.)

A contrary argument can be made when regulatory agencies discourage the use of fuel adjustment clauses. When utilities are not easily able to pass on increased fuel costs, higher expenses may lead to premature abandonment of gas and oil fuel plants (Carpenter, 1978, p. 3). This would accelerate the construction of more baseload capacity.

The most common types of fuel adjustment clauses in effect involve either assessing future costs and adjusting energy charges accordingly or incurring the fuel costs and adjusting rates to recover expenditures. The clauses may respond to the type of fuel purchased or its BTU content. In Sweden and England, utilities have indexed their rate levels by incorporating automatic fuel adjustment clauses and cost of living (or cost of construction) clauses (Mitchell and Acton, 1977, p. 5).

3.4.3 Flat Rates

Flat rates charge the same amount per kilowatt-hour to all users. A pure flat rate eliminates both demand and customer charges on the assumption that costs of serving large and small customers do not differ materially, so the charge per kilowatt-hour should be the same across all classes. The catch phrase is "a kilowatt-hour is a kilowatt-hour."

Flat rates are not cost based and offer no incentive for load management. On the average, it is concluded that flat rates would raise costs for industrial class users on the order of 35–65% (U.S. FEA, 1977a, p. 102). Initially, average electricity prices for small commercial and most residential customers would decrease, with reductions of 5–25% possible, whereas bills for larger customers would increase (U.S. FEA, 1977a, p. 102).

The long-run effects of implementing flat rates would include a decline in electricity consumption and a shift to oil and gas. The resulting lower load factors as electricity consumption declined (without a corresponding decline in peak period usage) would decrease the efficiency of utilities. The net effect on capacity is uncertain (U.S. FEA, 1977a, p. 112).

This particular rate structure also poses problems of equity and revenue stability. Presumably, flat rates would result in decreased kilowatt-hour sales to industrial customers, who actually represent the lowest costs of service to the utility. Costs would not be passed along to those incurring them but rather would be recouped in equal proportions from all classes of customers. This violates the fairness standard. Only if the costs of providing service did not vary across classes would the flat rates be fair. Flat rates also do not reflect marginal cost differences over time, as there is no variation in tariff with time of use. In this regard they cannot be said to give customers economically "correct" pricing signals.

On the positive side, flat rates would decrease monthly residential utility bills if usage were unchanged from current patterns. This, it is argued, would ease the rising costs of energy for low-income users. Further, the rates would tend to discourage wasteful consumption of electricity, if only because the tailblocks are higher than for declining rates now in effect.

3.4.4   Inverted Rates

Inverted rates are a mirror image of declining-block rates, as they price successive blocks of usage on an "inclining-block" basis. Under the name of inverted block rates, it is often proposed that industrial and commercial rates be kept lower than residential rates. The usual definition, however, is that the average price per kilowatt-hour paid for the initial block of residential usage would be less than the price per kilowatt-hour charged for all usage above that level. In this regard, inverted rates are similar to lifeline rates. It is claimed that long-run incremental costs support the concept of rate inversion because the tailblocks are responsible for plant additions and should therefore bear the costs of capacity additions. Noncost arguments, such as the rate's promotion of energy conservation and reduced charges to low-usage customers, have also been forwarded. These issues will be considered in turn.

The concept of inverted rates is often combined with a form of LRIC; the revenue requirement is allocated by one of the standard cost allocation methods (peak responsibility, noncoincident peak responsibility, or average and excess demand approach), but the tailblock residential rates are based on an estimate of LRIC. If the rate tends to produce excess revenue, as will be the case if future costs are rising, excess revenues are rebated in the initial block. The initial block will then be below average cost and constitute a form of lifeline rate.

Because the price of electricity increases with consumption under inverted rate proposals, large industrial users with large load factors will experience significant price increases. It is estimated that these increases will be on the order of 84% (U.S. FEA, 1977a, p. 103). If current estimates of elasticity of demand are reliable, these industrial users can be expected to reduce their consumption or change fuels if alternatives are available.

Small users will enjoy a cost reduction while large users will experience a cost increase. Total consumption will probably decline, and the higher electricity prices will cause increased usage of oil and gas for consumers overall. Industrial users may switch to alternative fuels, and it is probable that the net effect will be a decrease in oil and gas consumption in all sectors. Overall capacity requirements of utilities should also decrease.

Inverted rates may cause a reduction in revenue because of reduced usage and income from the traditional, large-user customers, for whom the cost of supplying electricity will be significantly less than the price they will be charged. Initially, these financially able customers will assume a greater proportion of the revenue burden, but as they leave the system for other energy sources, utility costs will be spread over the remaining (residential and commercial) users.

The arguments in favor of inverted rates on conservation grounds are founded on the proposition that electric utilities should price electricity so as to achieve zero energy growth. The ECAP Study did not question the validity of this objective but stated that it was appropriately the domain of a defined public policy. Rate structures, it was said, should not be distorted for this end. In support of this position, the report stated that "there is reason to believe that the market for scarce non-renewable resources will not misallocate them over time. Since resources of this type are mostly held by major corporations and governments, it could be rationalized that it is in their long-run best interests to maximize the current value of the resource over time" (Ontario Hydro, 1976f, p. 10). Continuing this argument, the report quotes Baumol and Oates (1975, p. 69):

> An ideal . . . market for our scarce resource can lead to current prices that reflect fully the social costs of consumption of the item. . . . If exhaustion of our petroleum reserves simply hastens the day when we will have to make use of solar energy which . . . will be very costly to process, the price of oil will rise as the date of substitution approaches, because of its rising opportunity cost; in a competitive market, this will be reflected as a higher current price.

Inverted demand rates have been tested by the U.S. Department of Energy in Vermont. The rate had some foundation in peak load pricing; it had no customer or energy charges, but only a demand charge reflecting the costs of increased capacity. The charges were $4.20 per kW for the first 3 kW; the next 3 kW at $7 each; and, finally, $8.40 for each kW over 6 kW.

The Vermont study concluded that the inverted demand rate did not work out well for either the customers or Green Mountain Power, the sponsoring utility. Customers lost interest toward the end of the year and the average customer's bill increased from 17–63.2% each month of the study (U.S. FEA, 1977b, p. 30). Because electric heating increases the load factor, the lowest average bill increases occurred during the winter heating months. The rate did not have any beneficial effect on the utility's system load, because there was no explicit incentive to shift consumption from peak hours, only to decrease maximum (instantaneous) demand. The report felt that the charges as originally set were higher than necessary and that customers could not adequately respond to the rate without rewiring their homes (e.g., using a lock-out switch to prevent the hot water heater and clothes dryer from being operated simultaneously) (U.S. FEA, 1977b, p. 31).

3.4.5 Ratchet Rates

Ratchet rates, also known as excess demand charges, are a surcharge levied per kilowatt when a residential customer's demand exceeds a specified level. The rates can either be stated as a flat surcharge per kilowatt—such as $1–2 per kilowatt of demand in excess of 15 or 25 kW, respectively—or as a percentage of the greatest billing demand incurred in the past year; that is, a 50% annual ratchet.

Some U.S. utilities offer the opportunity to have demand meters installed, which enables a customer to receive a discount on all demand greater than a fixed number of kilowatts. Ohio Edison has had such a rate in effect for 40 years where all kilowatt-hours in excess of 125 kWh per kW of demand are billed at a discounted charge. Although this is applied to declining-block rates, the principle can be used in conjunction with time-of-use pricing by offering the discount only in selected periods (e.g., summer or winter), or times of day. Unfortunately, when offered the option of installing demand meters, most customers decline for fear that their utility bills would rise. In the Ohio case, although customers could only benefit from having the meters installed, many elected not to do so. The Illinois Power Commission calculated that 20,000 of its customers would benefit from demand rates, yet only 23 of that number informed of this benefit by the utility decided to be served under the rates (U.S. DOE, 1979b, p. III-5).

W. W. Carpenter (1978, p. 24) of Ebasco Services credited the stabilization of load and energy growth for Central Vermont Public Service Corporation (a small, winter-peaking utility) to the use of the ratchet rate. The utility, it should be noted, also had a stored space heating rate, an insulation program, and controlled water heating.

The features and benefits of the rate were that (1) no additional cost of metering was necessary, (2) no additional billing costs or information were necessary beyond the prior 12 months' usage history, (3) no demand, only total usage was billed, (4) built-in "budget billing" accomplished a smooth transition from the previous rate to the current rate, and (4) a 2.4-to-1 seasonal price differential was imposed. This last feature was credited by management as significantly reducing the winter demand for electricity (Carpenter, 1978, p. 25).

The rate is notable for its simplicity. It also contains a highly important information component to consumers that helps them regulate their usage and compare utility costs. This feature alone, along with a seasonal price differential, involves minimal investment by a utility and could accomplish much in achieving the goal of load leveling.

3.4.6 Penalty Pricing

Penalty pricing is not properly considered a rate structure; rather, it is a means of penalizing and limiting excess demand. Instead of using circuit breakers when demand has reached a certain level, a penalty price goes into effect. There are few advocates of this scheme in the literature. In practice it is actually an alternative to interruptible service, a concept that has found much broader acceptance.

A related rate, called the limiter rate, allows small customers to install a demand limiter that interrupts supply whenever power drawn exceeds the setting of the limiter. Resetting is possible either manually or automatically. Rather than imposing a penalty, the limiter rate allows a discount for remaining within the demand limits.

3.4.7 Lifeline Rates

Lifeline rate schedules offer uniform low rates for the first several hundred kilowatt-hours of usage per month by residential customers. There is no change in revenues, as revenues lost because of the lower initial block are recouped elsewhere in the rate schedule. Proposals for lifeline rates vary as to whether revenues would be made up by higher residential or commercial and industrial rates. Other forms of proposed lifeline rates would require a means test for qualification.

It is argued that lifeline rates would alleviate the burden of rising energy costs to the poor, who, it is said, consume less energy than other residential customers. Studies alternately support and reject this hypothesis. It cannot be said that the poor necessarily consume small quantities of electricity or that small users of electricity are necessarily poor (U.S. FEA, 1977a, p. 77). Among the poor that would not be reached are those living in master-metered apartments or those consuming more than lifeline quantities of electricity. Low-income families who used other fuels instead of electricity would be discriminated against in a lifeline rate proposal. A U.S. survey indicated that lifeline rates would leave out more than half the poor in 14 states and more than 25% of the poor in another 25 states (Ontario Hydro, 1976f, p. 12). Others, however, have been more sanguine about the usefulness of lifeline rates based on experience in Los Angeles (Acton, 1980; Ray and Stevenson, 1980; Sullivan, 1979).

In economic terms, any decrease in price would tend to increase consumption, a result inimical to conservation. An alternative proposal to lifeline rates that would help the poor cope with increasing energy costs is an energy tax credit. Relief in this form would cover all energy sources and would more accurately identify qualifying households through the existing tax system.

3.4.8   Demand Energy Rates

Demand energy rates consist of a monthly charge for the first kilowatt or less, a monthly charge for each additional kilowatt of billing demand, and a monthly charge for all kilowatt-hours used. The billing demand is determined by a separate meter and is commonly calculated as the average kilowatt used during the 15-minute period of maximum use during the month. In effect, billing demand is based on each customer's own peak period usage.

The first step in the calculation of demand-energy charges is the determination of the total costs of the residential class, where total demand is the sum of maximum demands of each individual customer in the class. The calculation of system peak will be less than this total demand figure, as all customers' peaks are not coincident. Therefore, total demand is adjusted by what is known as a "diversity factor," defined as the ratio of the sum of each customer's maximum demand to the total class demand at system peak. The allocation of costs into the three components—customer, demand, and energy—would be as follows: total customer costs divided by the number of customers, total demand costs divided by the number of units of service (kW) adjusted by the diversity factor, and total energy costs divided by the number of units of energy (kWh).

Calculation of demand-energy rates using a marginal cost analysis yields different results. As before, the marginal cost of peak demand is the additional cost that would be incurred if demand at the customer level increased by one more unit at the time of system peak. The demand charge is again adjusted for diversity and a seasonal energy charge

TABLE 3.1.   A Comparison of Demand Energy Rates with Different Costing Assumptions

	Embedded Costs	Marginal Costs
Customer charge[a]	$8.73	$8.73
Monthly demand costs	$3.82/kW/month	$2.14/kW/month
Energy costs	2.16¢/kWh	Spring/Fall: 2.40 cents/kWh
		Summer: 3.15 cents/kWh
		Winter: 2.78 cents/kWh

Source: U.S. DOE, 1979d, pp. II-9, II-11.
[a]Assumed the same for both costing methods.

is included. A comparison of costs derived through a simulated model of utility costs is set forth in Table 3.1.

The Public Service Company of Colorado offers demand-energy rates optionally to residential customers with electric space heating. Most new homes in the service area are equipped with demand-limiting devices that shut off low-priority circuits when the customer's demand begins to build (U.S. DOE, 1979b, p. III-1-2).

One type of rate related to a demand energy rate has been tested in Vermont. Called a "peak kilowatt demand rate," it has been described as a "peak-shaving" rate. There was a flat energy charge of 1 cent per kWh on all energy. There was also a capacity charge of $8.40 per kilowatt that was applied only to demand during the peak load hours of 9 A.M. to noon. One customer beat the rate by overheating his home prior to 9 A.M. and then shutting off the main breaker until after noon. The result was a zero demand charge and total consumption billed at 1 cent per kilowatt-hour.

The rate did not meet a high degree of acceptance either by the company or its customers. If customers were able to keep consumption to a minimum during peak hours, savings would result. For Green Mountain Power of Vermont the rate was unacceptable for at least three reasons: (1) the rate would require fine adjustment to eliminate the high savers and high losers; (2) the rate created sudden high increases in load immediately following peak hours, which created a new system peak; and (3) the metering hardware required for a time-of-day demand charge rate is very expensive compared to a time-of-day kilowatt-hour rate (U.S. FEA, 1977b, p. 24).

The problem with demand rates is that once the high demand for the month is established, accidentally or otherwise, there is little incentive to conserve, nor is much opportunity left for savings.

3.4.9   Contract Rates

A contract rate enables a customer to decide on his maximum level of demand in advance for which he is billed, whether or not the power is consumed. Also known as "subscribed demand" or "take or pay," the rates reduce forecasting errors for utilities, or at least shift responsibility for forecasting to the customers.

Contract rates were also tested in Vermont. The rate in effect had no energy charge, but the residential customer contracted for a given number of kilowatts at $4.25 per kilowatt. There was also a penalty charge of $8 per kilowatt for all kilowatts in excess of the subscribed rates.

Some customers did well under the rate schedule, and others could not adjust at all. The study found that group daily load curves exhibited frequent high and sharp peaks occurring at random. Customer satisfaction varied. The contract rate customers, like the demand rate customers, had trouble with infrequently used, high-demand appliances like clothes dryers. The study concluded that automatic demand control equipment would be necessary if the contract rate were to be acceptable to all customers (U.S. FEA, 1977b, p. 32).

3.4.10   Interruptible Power Contracts

Industrial Interruptible Load Management Rates

Interruptible power service is a longstanding and acceptable practice among electric utilities. Although the number of subscribers to such rates is typically small—usually one to three controlled customers per utility—the rates represent substantial savings to those industries able to curtail their power on notice. The most common interruptible industrial processes include arc and alloy furnaces, steel mills, mineral crushing (including coal), rolling mills, induction heating applications, and chemical processes (U.S. DOE, 1979b, p. V-7).

Notification procedures differ between utilities; some give as little as 15 minutes' advance warning and others give 24 hours. A distinction is also made as to who controls the interruptible load. In most instances the customer controls his own load. The conditions and duration of load curtailment vary widely. There may be regularly scheduled curtailments, maximum established outage periods, or interruptible service only under defined conditions. Different types of penalties are imposed for failure to comply, and occasionally the utility cuts the customer's interruptible power if a request is not executed. The most common form of rate is a discount on kilowatts or kilowatt-hours of interruptible power used in the form of a credit to the regular bill.

The pros and cons of interruptible rates can be summarized as follows (Ontario Hydro, 1976i, pp. 2–5):

1. Interruptible load forecasting is *subject to oscillation and uncertainty* due to business cycle fluctuations; so future load patterns may differ considerably from the past.

2. Interruptible loads constitute a form of *available capacity;* if the total system reserve level, and therefore the LOLP, were to remain unchanged, then a utility would need less reserve. Lower generation capacity required from use of interruptible supplies would result in substantial cost savings to utilities.

3. It should be recognized that the more interruptible power sold, the higher the probability of cutting off customers. This also applies if the general level of system reserve is lowered.

4. The savings accruing to the utility from the adoption of these rates would be realized for all customers (i.e., fixed charges today are lower than if all load had been forecast as firm). Because both planned system capacity and planned reserve would both be higher without interruptible power, there are savings to be gained whether cuts are actually made or not.

5. If excess capacity exists, the reliability of interruptible power should be high; *the incentive to subscribe to interruptible power service would decrease as the risk of interruption increased.*

6. If firms left interruptible power service for firm service on short notice or, conversely, chose to subscribe to interruptible power on short notice, the stability of the system would be strained.

Ontario Hydro has approximately 30 interruptible contracts in effect. There have been very few interruptions in service. This is because of Hydro's excess capacity and the corporation's practice of cutting export of electricity when a loss of generation occurs rather than cutting its interruptible customers. Compared with these very lenient circumstances, another user of interruptible contracts, the Bonneville Power Administration in the Pacific Northwest, imposes very stringent conditions on its subscribers. This is probably so because it is largely a hydro system and the power source depends on seasonal conditions.

The Norwegian Water Resources and Electricity Board (an all-hydro system) has contingency plans for a dry year by supplying some large industrial customers (aluminum and hydrogen producers) with power under interruptible contracts that permit interruption for several months. This is priced at 75% of the secure power price (Mitchell and Acton, 1977, p. 21).

The ECAP Study recommends that interruptible power contracts be sold via an auction, which would better equilibrate supply and demand and invite more customer participation. The amount of interruptible power offered for sale would be related to varying probabilities of interruption. This information would be provided to the customer. Each customer would conduct his own cost/benefit analysis and bid a price. His position in the total supply of interruptible power would be determined by the bid price, with the lowest-priced contracts subject to the earliest curtailment. The current system employed by Ontario Hydro divides cost savings 50/50 between firm and interruptible customers.

Residential Interruptible Contract Rates

A commonly proposed residential form of an interruptible rate consists of a discount for a specific type and number of interruptions (with an upper limit on the number of interruptions and the hours per interruption). Discounts increase with the number and duration of interruptions. In addition, customers served at lower voltages would bear higher costs because of the additional distribution losses involved in serving them. Equipment costs would be added to the fuel and capacity costs. The optimum rate structure would be calculated as the reduction in the present value of revenue requirements realized, translated to a per-customer discount that exactly exhausts the utility's savings resulting from the program. This approach has been implemented by the Wisconsin Electric Power Company and adopted by other utilities (U.S. DOE, 1979d, p. II-18). This approach to rate design must account for the "lumpiness" of investment, the phased-in nature of the rate likely to be necessary, and the year-to-year variation in savings due to generation mix, capacity availability, and system load levels. Use of present-value calculations and a long time horizon helps to smooth out these investment timing differences. (Use of these long-term approaches, however, may conflict with the necessary calculation of revenue requirements and cost allocations on an annual basis.)

Load management discounts calculated for a "typical" Wisconsin utility (a synthetic utility created by using aggregated statewide data) are listed in Table 3.2.

The precise level of discount will vary by utility, of course, and will depend on the assumptions made concerning its interruption strategy. If interruptions are staggered to reduce the bounce-back of usage around peaks, the peak savings will be reduced. For example, individual users might be cycled off for the first, second, third, or fourth quarter of an hour around peak. This permits system load to be reduced for a longer period of time than any individual customer might tolerate. A blanket interruption would be more efficient but less acceptable to customers.

Discounts might also be manipulated initially as an inducement to cycling strategies then returned to a purely cost-based structure later once acceptance has been ensured.

Green Mountain Power also studied the effects of interruptible rates on its residential customers. The rate took the form of a $5.75 monthly credit on the standard residential rate in exchange for the disconnection of the customers' electric hot water heater during the periods 9 A.M. to noon and 5 P.M. to 7 P.M. The customers on the interruptible rate exhibited noticeable changes in their daily load curves. There was a consis-

TABLE 3.2.   "Typical" Load Management Discounts with Interruptible Contract Rates

Electric Water Heating Controls		Central Air Conditioning Controls	
High Saturation	Low Saturation	High Saturation	Low Saturation
$1.60/mo	$3.38/mo	$0.85/mo	$0.97/mo

Industrial Interruptible[a]	
4 hour	12 hour
$2.84/kW/mo	$4.16/kW/mo

Source: U.S. DOE, 1979d, p. II-20.

[a]Equipment costs for industrial load control are small on a per kilowatt basis of controlled load.

tent drop in load during the mornings and a sudden increase at noon. The evening effect was much less noticeable. The utility was sufficiently optimistic about the rate to study it further (U.S. FEA, 1977b).

3.4.11   Summary and Conclusions

Marginal cost pricing is generally taken to be synonymous with time-of-use pricing. Rates with some cost basis tend to have a time-of-use component, whether seasonal or diurnal. Time-of-use rates on a seasonal basis can be implemented very easily as no additional metering is required. They are particularly appropriate in hydro systems, which have relatively constant daily generating costs. Billing information incorporating past consumption information is another simple yet effective means of modifying customers' behavior without a radical change of tariff. Overall, the non-cost-based rates seem the least acceptable and pose the greatest equity problems. These include flat rates and lifeline rates. Declining-block rates, which formerly had some cost basis, contribute to the overbuilding problem and foster continuing rate increases. Contract rates and interruptible power rates provide a means by which the customer can choose a rate structure most consistent with his needs. Contract rates are particularly appealing insofar as they stabilize forecasting, match demand with cost, and make electricity consumption decisions on the part of customers more calculated and deliberate. Both contract rates and interruptible power rates require some sort of mechanical device for enforcement. All the rates reviewed can be based on embedded costs and are therefore within reach of today's utilities.

4

ELECTRICITY PRICING: LOAD MANAGEMENT

The central issue in deciding whether to implement a load management program rests with the relative effectiveness of rates versus mechanical devices as a means of controlling consumption. Load management is a means of moderating short-term fluctuations in demand. The long-term effects of a more level load curve are uncertain; the short-term effects include increased system reliability, greater available capacity, and reduced fuel costs.

4.1 LOAD MANAGEMENT STRATEGIES

4.1.1 Time-of-Use Rates

Time-of-use rates are usually formulated as follows: Residential customers are on a schedule that includes a customer charge and a separate time-varying energy (kWh) charge, but no separate demand (kW) charge. The energy (kWh) charge contains no usage blocks, but is lower during the off-peak period. The peak-period charge includes all the capacity costs imposed by the residential class plus the larger share of the energy costs imposed by such users on the system. The off-peak charge consists only of the remaining energy costs. The ratio of energy components of the peak and off-peak period prices is proportional to the ratio of incremental system running costs (system lambda) between peak and off-peak periods. Commercial and industrial rates include separate customer, energy, and demand charges. Energy charges vary by time of day in proportion to incremental running costs, and from class to class to reflect the costs of serving customers of different voltage requirements.

In this case, costs of capacity are assigned wholly to peak periods. "This marginal cost approach is neither a 'long-run incremental cost' approach nor an 'average cost' approach. . . . The former attempts to project the future costs of supplying electricity, while the latter tends to assign capacity costs to off peak as well as peak periods" (U.S. FEA, 1977a, pp. 74–75).

Time-of-use rates should result in lower utility bills for most residential and industrial customers. Commercial-class customers, however, might experience an increase as the commercial class is generally responsible for a larger share of peak demand than overall kilowatt-hour sales. Customers will also have to bear the cost of additional metering. The rates would not increase for industrial class users because their share of system kilowatt demand is less than their share of total kWh sales.

Insofar as customer costs under time-of-use rates are reduced for the residential and industrial class of customer, it might be observed that the rates will increase electricity costs, although the preliminary data available from the Wisconsin experiment (reviewed elsewhere in this book) indicated that overall consumption, as well as peak-period consumption, was reduced. In contrast, the Ohio rate experiment found that total consumption for the experimental group did not differ significantly from that of the control group (U.S. DOE, 1979e, p. vii). Lower consumption will correspondingly reduce capacity requirements.

Because time-of-use costs are felt to track utility costs more accurately than other cost structures, they should provide adequate revenue protection and stability.

As to the fairness of time-of-use rates, the standard declining-block rates actually subsidized on-peak users because rates were invariant with time of use. Elimination of this implicit subsidy should increase fairness to all classes of consumers to the extent that usage reflects costs imposed on the system. The critical issues under fairness are whether large users impose higher costs, and who is to be responsible for the baseload capacity costs during system peaking; these must still be resolved.

Costs to the utility under time-of-use rates will be increased because of the costs of metering individual households, changes in billing practices, reformulation of data processing specifications, as well as the experimentation that must necessarily precede any conversion of rates. Customer education and other mechanical load management techniques, if used in conjunction with time-of-use rates, will also be added to the utility's cost of service, ultimately to be passed on to consumers. The time required for the changeover is estimated to take a number of years once the initial decision has been made, primarily because of the low availability of time-of-use meters. Because of these costs in administration, equipment, and time, implementation of time-of-use rates imposes significant dollar costs on utilities that in each case must be weighed against potential benefits.

Introduction of time-of-use rates on a voluntary basis is a good means of experimenting with load management rates on a limited scale. The results might prove deceptive because customers availing themselves of the rates would be those most likely to have high off-peak consumption in the first place. One suggested incentive to make optional time-of-day rates of interest to customers is to provide (or confront) those customers who are not on time-of-day rates with an inverted rate in which the end block equals the seasonal marginal cost per kilowatt during daily peak hours (Wisconsin PSC, 1979f, p. 19).

In a cost/benefit analysis, the savings to individual customers represent cost savings passed along from the utility. A test of the rates in Vermont found that the average customer saved approximately 15% on his monthly bill; with the highest savings equaling 25.5% of the regular (declining-block) bill. (The Vermont rate had a daytime–nighttime block differential of 4.5:1.) The percentage of saving per month was slightly lower during the winter months because of the increased winter heating load (U.S. FEA, 1977b, p. 25). The Vermont study concluded that the time-differentiated rate clearly demonstrated a potential for improving the system daily load factor and reducing energy costs through more efficient use of generating units with lower energy costs (U.S. FEA, 1977b, p. 26).

Metering

The cost of metering for time-of-day rates is a significant obstacle to the commitment of utilities to load management programs. The metering costs increase with the complexity of the rate structure; for example, the cost of metering customers for demand charges is greater than metering equipment for time-of-day rates. Meters are currently available that can incorporate suggested rate modifications, although some lead time is required before demand can be met.

A comparison of single-rate, optional two-rate, and full-scale two-rate metering equipment was made in Ontario, projected over the period 1978–2000. Depending on the estimated sensitivity of customers to time-of-use rates, and hence the value of cost savings, the net savings through the use of optional two-rate metering was estimated to lie between $0.5 and $122 million (1978). "Full implementation of two-rate metering may increase costs by as much as $35 million (1978), or result in a net saving of up to $638 million (1978)" (Ontario Hydro, 1976h, p. 103). Clearly, estimates displaying this much variation are not very helpful to planners. The uncertainty is due to conflicting estimates of the price elasticity of electricity, which, to have any kind of validity, must be measured separately for each utility service area.

There is some foreign experience with two-rate metering, although there is none in Canada and only very little in the United States. Two-rate, time-of-day metering has been available in England and Wales for nearly a decade, and is widely used in Switzerland, the Netherlands, and France. West Germany has made a significant improvement in load factors by the use of load control and time-of-day metering. The rates are optional in Belgium (Ontario Hydro, 1976h, p. 103).

The cost of meters has fallen dramatically; the ECAP Study found that by November 1978, the price was about $79 (U.S.). Installation of time-of-use meters would also require the write-off of existing equipment, which utilities would perceive as an accounting loss.

A number of recent trends in utility billing also present obstacles to implementation of time-of-use rates. The first is the practice of longer meter-reading intervals as a means of cutting costs. Where billing information is not presented rapidly, the feedback effect is diminished. Inclusion of water bills with electricity bills also clouds the effects of changes in electricity consumption. Users' preference for bills averaged over the year would also have to be overcome. Some fear that the institution of time-of-use rates would introduce "a whole new dimension of residential customer complaints" (O.E.B., 1979, p. 173). Finally, utilities would have to face the increasing incidence of meter tampering and attempts at "rate beating" as customers became more aware of billing practices.

4.1.2 Bans and Restrictions

The most notable and successful application of an ordinance restricting electricity consumption occurred in Los Angeles during the 1973 embargo. Los Angeles utilities, particularly dependent on supplies of oil and natural gas, faced serious disruptions of service unless demand was cut. The city chose to proceed by way of an ordinance requiring customers to cut their consumption of electricity (over the previous year) by 10% for residential and industrial customers and 20% for commercial customers. A second phase was also designed (although never invoked) that would have cut residential use by 12% and commercial use 33% over the corresponding billing period a year before. The penalty was a 50% surcharge on the *entire* bill for the first period violation, and cutoff of service for subsequent violations (Acton and Mowill, 1976, pp. 2–3).

There was an immediate response to the enactment. In the first 11 days, electricity demand fell 14.9% over the period a year earlier (1972). Total sales were off by 20% during the first two months the ordinance was in effect and, in May 1975, over a year after the ordinance was suspended, total sales were below 1973 levels by 8%. Residential and commercial sales fell rapidly (18% and 28% below the 10% and 20% required by the ordinance) (Acton and Mowill, 1976, p. 5).

Changes in lighting accounted for most of the reduction at the individual level; most commercial establishments met their 20% target with changes in lighting alone (Acton and Mowill, 1976, p. 7). The Rand report studying the effects of the ordinance attributed its success in part to the setting of conservation targets that permitted a great deal of flexibility in conservation measures adopted by individual establishments. "This flexibility in individual response seems to have been very important in the widespread compliance that was observed" (Acton and Mowill, 1976, p. 18).

The use of the 50% surcharge (which was never imposed in practice) was one alternative chosen over the use of rolling blackouts and limitation of operation by businesses to 50 hours per week. In developing the ordinance a consensus was reached among business, labor, and government leaders that a percentage curtailment was a more attractive means of voluntarily meeting the shortage (Acton and Mowill, 1976, pp. 18–19).

The policy seems to have had a lasting effect on the pattern of consumption, not only causing a one-time reduction but also slowing the rate of growth. It was concluded that these long-run effects may be due to the general appeal to conserve as well as greater price awareness. The ordinance called attention to the financial savings that result from a reduction in consumption. This has since been reinforced by higher rates. "If the estimated elasticities reported from national cross-sectional data are used, they generally imply that price would have had to increase by more than 100 percent to achieve the 17 percent reduction in demand observed immediately after the enactment of the ordinance" (Acton and Mowill, 1976, p. 21).

As to whether the approach is transferrable to other utilities, it is important to recall that the ban was the result of recommendations of a broadly based committee of business, civic, and labor leaders. Further, the penalties were never actually invoked. If they were, the Rand report notes, it would create a significant administrative burden. Without the penalties, 10% of the residential customers appealed for an adjustment in their percentage, whereas less than 2% of the commercial and industrial customers appealed. "Had the penalties of the ordinance actually been levied, it is possible that the number of appeals would have risen to a level that would have completely overburdened the system" (Acton and Mowill, 1976, p. 22).

The success of the restriction is remarkable because of the level of voluntary compliance; it also underscores the amount of excess consumption that can be trimmed with relatively little effort.

Other forms of restriction include outright bans on certain end uses of electricity, such as air conditioning or display lighting.

4.1.3   Remote Load Control Devices

Remote load control devices are utility-activated devices triggered by either radio signals, high-frequency impulses over a power line, low-frequency ripple signals over a power line, or telephone pulses. Load devices are a sophisticated mechanism for allowing direct utility intervention in the use of electricity. They provide greater flexibility in load management than time-of-use rates because they allow the utility to reduce or switch loads automatically whenever necessary. They are particularly effective at the residential level in conjunction with water heaters because service interruptions appear not to be noticed by customers.

Load control devices are attractive for a number of reasons (NARUC, 1977, p. 55):

> First, if loads tend to be sharply peaking because of irregular events such as weather, then load controls may be more effective than time-differentiated rates. Second, if customer costs for storage devices or other equipment to respond to time-differentiated rates are large, then load controls may be more cost-effective. Third, load controls may be necessary to achieve load management objectives . . . if customers respond weakly to cost-based price incentives. . . . Fourth, . . . direct load controls may be a cost-effective means of dampening peak load growth . . . . [and] could forestall adding new generation capacity. . . . Further, the use of controls allows a utility to maintain a given level of reliability with less capacity by selectively reducing its level of service to particular customers at specific times. . . . Load controls have a certainty not found in time-differentiated pricing approaches. . . . [and] controls can be applied exactly as needed, more closely approximating the economic efficiency results of *short-run* marginal cost pricing.

There are a number of difficulties associated with remote load controls. The first is the criticism that they violate "consumer sovereignty." This can be avoided if the proper compensation or incentives are offered for allowing the remote cycling of certain loads.

Another consideration is that cycling of appliances will increase LOLP during the shoulder peak, relative to the LOLP during the peak period, so the monthly credit for accepting a load management device should be adjusted for increasing shoulder-period demand costs (Cicchetti and Reinbergs, 1979, p. 235). It also has been suggested that customers merely be offered a small incentive to have their hot water heater thermostats reset to a slightly lower temperature, thus saving the utility the cost of a load management device (Cicchetti and Reinbergs, 1979, p. 235).

One method for "beating" remote load control is by raising the temperature of a water heater so when the unit is cycled off, the effect is diminished. Customers might eventually respond by buying oversized units to avoid any inconvenience, so if the credit was not tied to appliance size, little benefit would result. As long as the device is utility activated, customers can be expected to try to avoid the consequences. If, however, devices are controlled by the customer directly, the incentive is stronger to maximize his savings.

Remote load control devices are a form of thermal energy storage, that is, a means of storing off-peak electric energy for thermal applications during the peak load hours. Besides being relatively unnoticeable to customers, interruptible service has the further attraction of not requiring any direct investment by the customer. The utility-installed control device is actually a substitute for a bulkier water storage unit that might be purchased by the customer to benefit from time-of-use rates.

> The British Electricity Council study of household load curves suggests that the installation of storage heating has typically resulted in shifts of 0.8 to 1.4 kW in a household's morning and early afternoon demand on an average day in December and January. On cold days, the shifts are large, ranging up to 5 kW per home. The study implies that on an average winter day storage heating shifts a total of 2000 to 3000 MW out of the daytime system load curve and a smaller amount out of the evening load into the nighttime load; and on a cold day, the shift can be 9000 to 10,000 MW (note: Not all this amount represents a shift from peak to off-peak hours, because part of it is due to the substitution of electric heating for systems fired by coal or fuel oil) [Mitchell et al., 1977, pp. 49–50].

The effects of storage heating are dramatically illustrated in Figure 4.1.

Interruptible service is best used in conjunction with time-of-use rates. The rate would be in the form of a monthly discount for all customers who permit interruption of their hot water heaters for periods up to, for example, four hours a day. The rate could take the form of a contract for a certain time during which the customer would be guaranteed that both the rates and length of the peak period would remain in effect.

Time-of-use rates in effect during off-peak periods do not entirely correspond to the system valley. The system peak might be 12–14 hours, while the optimal storage time might be 8 hours. This would give rise to two problems: first, the threat of thermal storage actually adding to the system peak, and, second, the relative price paid for off-peak energy being too low. Potentially, the entire load might shift to the first hour of the off-peak period (known as the *payback phenomenon*), although staggering the peak periods for different customers would alleviate this problem.

The second problem can be visualized with reference to Figure 4.2. As a system valley is filled, the marginal costs of energy increase with load.

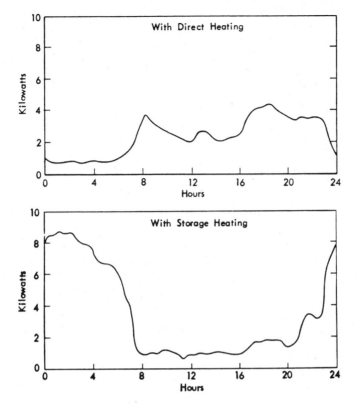

FIGURE 4.1. Daily pattern of electricity consumption in two types of English homes with different methods of space heating (at a winter weekday temperature of 32°F, normal insulation). (Source: Mitchell and Acton, 1977, p. 45.) © The Rand Corporation. Reprinted with permission.

Related problems arising from widespread use of remote load control devices involve adjustments to sustain capacity, potential unbalancing of the local grid caused by the large loads required by storage systems, and the additional costs necessary to install and inspect the devices (Nelson, 1976, pp. 41–42).

4.1.4 Nonremote Load Control Devices

Nonremote control devices are self-contained units located on the customer's property that limit his demand for electricity in some way. The most common devices include:

1. A clock that triggers a relay at preselected times during the day or week.

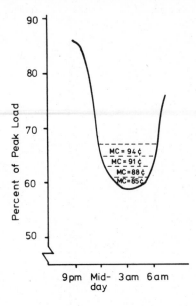

FIGURE 4.2. Hypothetical load curve showing change in marginal cost (MC) as system's diurnal valley is filled. (Source: Nelson, Argonne National Laboratory, 1976, p. 42.)

2. A circuit breaker (or fuse) that turns off loads when current passing through it exceeds its rated value.

3. A thermostat that switches loads according to predetermined temperature levels.

4. An interlock that allows one or several but not all major electric appliances in a home to be on at one time (NARUC, 1977, p. 56).

Forms of demand limiters are already in use by commercial and industrial customers. These monitor the total power used by the customer and disconnect loads according to a predetermined sequence established by the customer.

All the methods would have little effect if not used in conjunction with time-of-use rates. Warning devices that signal customers when the utility is operating "on-peak" are only meaningful if a price differential takes effect at that time. Similarly, with interlocking devices there must be some specific economic incentive available for the customer to voluntarily limit his electricity consumption.

A form of demand limiter is in widespread use in France. It is similar to a circuit breaker and the customer purchases it on a subscribed demand basis. The more he pays, the higher his simultaneous level of demand can be without activating the device. The demand charge normally includes a portion of the utility's capacity costs. There is no peak/off-peak rate differential, as the overall level of demand is known with relative certainty to the utility.

4.1.5   Energy Conservation

Conservation is a load management strategy insofar as it implies increased appliance efficiency as well as reduced usage. Conservation implies more of a reduction in total energy consumption than merely a reduction in peak period energy use. An overall reduction in demand with no change in peak demand implies a worsening of the system load factor, that is, a wider gap between average and peak consumption. In a peak critical system this would neither save utility fuel generation costs nor lead to a reduction in planned capacity expansion because actual and projected growth of peak demand would have to be met.

As suggested before, if electricity is priced below its marginal social cost, the incentive exists to consume more, not less. A corollary to this is that because electricity rates are being maintained below marginal costs, the private incentive to adopt electricity-conserving changes in technology is also less than optimal (Efford, 1978, p. 62).

It has been assumed throughout that marginal cost pricing will lead to a rise in the price of electricity. A rise in the price of electricity is not, however, inconsistent with lower average monthly utility bills for customers, as was shown to be the case under forms of time-of-use pricing. An argument made against higher electricity prices is that they will accelerate the consumption of fossil fuels, and that reducing the consumption of fossil fuels should take precedence over electricity conservation (O.E.B., 1979, p. 29). Besides the obvious response that conservation of electricity will reduce the amount of fossil fuels burned in power generation, conservation is not fuel substitution and does not imply such.

One approach proposed has been called Turvey's Short-Run Pricing Rule. The author of this theory

> explained that if there is no possibility of a shortage of capacity there is no reason on grounds of economic efficiency for conserving electricity. . . . In such a case, as much electricity as possible should be sold, provided that customers pay the marginal fuel costs. . . . He argued too that prices should be reduced to stimulate demand and that it would be doubly wrong to follow the mistake of over building capacity by the mistake of not using it [O.E.B., 1979, pp. 48–49].

This rule makes economic sense as long as the additional demand does not itself stimulate more overbuilding. The exportation of electric power is an increasing trend in Canada and may change the nature of the industry itself. Net exports have grown from near zero in the 1960s to 6% of total production in the 1970s. For example, Ontario Hydro exports by 1978 equaled nearly 12% of its generating capacity (*The Financial Post*, 1980, p. 31).

4.1.6 The Effect of Load Management Strategies on Utilities

There are few data available on the effects of load management strategies on utility costs: first, because the only programs in North America are either very small scale or experimental; and, second, because the nature of generation costs vary so widely among utilities as to make cost comparisons difficult.

One U.S. study took data from two "typical" utility systems, a primarily coal-based system with large industrial loads, and a primarily oil-based, summer-peaking system with predominantly residential loads. The study simulated the effects of load management strategies and calculated the savings in fuel and operating expenses. The experiment examined the effects of changes in load shapes; seasonal shifts in load were not included. The results of the sensitivity studies revealed that:

1. Fuel costs were significantly reduced for the coal-burning utility by reducing peak loads, almost regardless of the degree of energy conservation.

2. Fuel costs were significantly reduced for the oil-burning utility through energy conservation, almost regardless of the change in peak load.

3. Pure conservation measures (reduction of energy supplied on-peak without a corresponding increase in off-peak) had an adverse effect on both utilities' net operating revenues, defined here as revenues minus operating costs, because reductions in revenues exceeded cost reductions.

4. Fuel costs declined with . . . increasing load factor, but approached a lower limit well before load factors reached their theoretical limits (U.S. FEA, 1976b, p. 5).

Four types of load management strategies were tested and only two were found justifiable in terms of cost. Time-of-day rates for industrial-class customers of the coal-burning utility were attractive; time-of-day rates for commercial-class customers of the oil-burning utility were also found to be justifiable. Two other strategies—remote water heater control and thermal insulation—fell short of being justified solely on the basis of short-term operation cost reductions. However, potential long-term savings from generation investment deferral are substantial enough to justify remotely controlled water heaters (U.S. FEA, 1976b, p. 6).

Figure 4.3 sets forth a parametric analysis of load shape modifications under different assumed reductions in overall energy consumption and peak load changes for the oil- and coal-burning utilities under study. The cases are identified by two parameters, the percentage change in peak load and the percentage change in energy generated. The reference case is 0.0, 0.0. For the oil-burning utility, increased reductions in peak load, with energy held constant, reach a leveling off in cost savings although

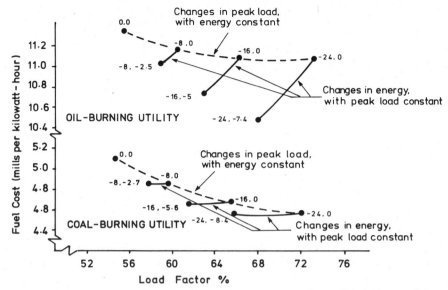

FIGURE 4.3. Average fuel costs vs. load factor, parametric analysis of load shape modifications. (Source: U.S. FEA, 1976b, p. 35.)

the load factor continues to improve. This means that once a certain level of peak has been shaved, cost savings are fairly constant thereafter. The other case, where peak load is held constant and energy use drops, shows a dramatically *increasing* reduction in per kilowatt-hour fuel costs as the level of energy conservation rises. (This reflects the utilities' increasing short-run marginal fuel costs.) The second example, that of a coal-burning utility, displays very rapidly increasing cost savings as load management is implemented without overall energy consumption changing.

These findings can give a general indication of how load management strategies influence utility costs, but individual systems will behave differently. The conclusions as to which strategies are cost effective and which are not should be treated with more care, as the increase in fuel costs since the study was conducted (using 1974 data) has made many options cost effective that were not previously.

The study recommended that the priority of implementation for time-of-day rates should be from large customers toward smaller ones. Before embarking on a given load management strategy, utilities should be required to show estimates of potential fuel cost savings under these rate structures. The study could not conclude whether the utility should include in its estimate of cost savings the changes in future expansion plans, as these data were more elusive and difficult to quantify. If the strategies could be justified on the basis of fuel savings alone, that would be a minimum condition of cost effectiveness.

The Arkansas price experiment conducted by Arkansas Power and Light calculated that if all customers were on time-of-day rates, costs for 1976 would have been reduced by $20 million. If these costs were returned to customers, less metering expenses, it would amount to $53.28 per customer (U.S. DOE, 1979?, p. VII-2).

4.1.7 Summary and Conclusions

Load management includes any technique that attempts to level a utility or household load curve. Direct load management via either utility-controlled or customer-controlled devices is held by proponents to be superior to time-of-day rates because capacity savings can be calculated with greater certainty. Use of time-of-day rates requires knowledge of the elasticities of demand of customers in order to project the effects of the rates. This information is difficult to obtain and not always reliable. Successful load management should result in use of more baseload and less cycling and peaking capacity. If the utility bases its expansion decisions on the maintenance of a certain level of system reliability, plant expansion can be delayed because of the improved load factors. The greater the savings from fuel and capacity investment, the more quickly other forms of load management become cost justified. Individual utility load management programs must weigh potential cost savings from fuel against the need for increased baseload capacity to serve flatter loads. Many environmental groups appear to reject time-differentiated rates on the ground that they promote increased off-peak usage and thereby hasten the growth of nuclear capacity at the expense of solar power. Load management programs in general represent a centralization of decision-making about the use and supply of energy and therefore take some of these decisions away from the consumer (Cicchetti and Reinbergs, 1979, p. 233). The relative merits of load control versus overall conservation of electricity should be a public policy issue.

4.2 CUSTOMER RESPONSE TO LOAD MANAGEMENT RATES

There are few hard data available on the effects of load management strategies, and what are available should not be treated as generalizable to all utilities. This caveat applies particularly to analyses of price elasticity of demand for electricity. Before the U.S. Department of Energy undertook the vast Electric Utility Rate Demonstration Program evaluating various innovative pricing and load management techniques, the available information was based on aggregated data and theoretical models. Results of some theoretical models are presented here along with some of the results of the Rate Demonstration Program.

4.2.1 Residential Customer Response

Customer Understanding and Acceptance

Overall, customers who have been put on time-of-use rates on a voluntary basis, at least under test conditions, elect to continue and respond
favorably to questions concerning the rates. Consumers in the Connecticut time-of-use pricing experiment by and large believed that the test
rates gave them the opportunity to save on their electricity bills by shifting use (Connecticut PUCA, 1977). Customers who were put on the
rates with little or no advance information (in an effort to counter any
possible Hawthorne effect) often resented having been subjected to the
experiment. The Arkansas study sent test participants only a brief description of the experiment, which was mandatory for those selected.
The three rates tested were: (1) a time-of-day rate with a 6 : 1 summer on-
to off-peak price differential; (2) a seasonal rate with a 3 : 1 on- to off-peak
price differential; and (3) a second seasonal rate with a 2 : 1 differential.
The comments made by the time-of-day group were the most favorable.
The negative comments reflected resentment at having been subjected
to the rates without choice. The customers subjected to the 3 : 1 seasonal
rate voiced the most outrage and felt, probably rightly, that the excess
should be returned to them. The 2 : 1 seasonal rate participants still felt
that the rate was too high, with unfavorable comments dominating, such
as: "something should be done about this or the government should take
over." Of the respondents to the postexperimental questionnaire, 25%
said they would like to keep the current experimental rate, while 75%
desired the old rate (these included responses by respondents from the
control group, who indicated a preference against the current experimental rate even though they were not on it) (U.S. DOE, 1979?, p. IV-
17).

The Connecticut experiment conducted an interview to determine
attitudes toward electricity prices. Seventy-two percent of the 199 customers surveyed felt that the cost of electricity was "too high." An overwhelming majority (82%) believed that the situation regarding prices
would not improve and that the cost of electricity would increase in the
next several years (Connecticut PUCA, 1977, pp. 26–27).

The conclusion to be drawn is that any rate reform must be accompanied by a major customer education program detailing the rate structure and how customers might benefit. A complete customer education
program should include information on available load management
equipment as well. "Experience in Europe indicates the importance of
disseminating information [on timing and storage devices], for those
utilities that were most active in promoting devices to take advantage of
the rates showed considerably greater load-factor improvement than
those that were less active" (Nelson, 1976, p. 27).

Demand Response

ELASTICITY OF DEMAND

The demand function traces the relationship between market prices and quantity demanded by consumers; "price elasticity of demand" refers to the effect of incrementally small changes in the price and the corresponding changes in the quantity demanded. This describes the sensitivity of customers' demand for electricity at different prices and is the fundamental issue in formulating rates. In the power industry demand refers to time-differentiated kilowatts or kilowatt-hours. Thus, there are two measures of elasticity: one for power at any given time—kilowatts, which can be referred to as *peak elasticity*, and the other for total energy consumed—kilowatt-hours, *elasticity of demand for energy*.

Price elasticity of demand for energy is the ratio of the percentage change in consumption (measured in kilowatt-hours) to a given percentage change in the price of electricity (measured in cents or mills per kilowatt-hour). Because an increase in price is associated with a decrease in consumption, price elasticity is usually negative.

Elastic demand is normally defined by economists as greater (i.e., more negative) than -1.00; conversely, inelastic demand lies between 0 and -1.00. Inelastic demand implies that the percentage change in consumption is less than the percentage change in price. Table 4.1 sets forth elasticities for the residential, commercial, and industrial sectors on both a long- and short-term basis.*

The conclusions derived thus far from the various studies indicate that demand for electricity is more elastic in the long run, when the appliance stock is allowed to vary, and less elastic in the short run.

A further indicator of elasticity is the cross-elasticity of demand between peak and off-peak periods. In most cases these figures must be derived on a utility-by-utility basis by experimental means before rates can be established and revenues calculated. One important conclusion of many studies of the price elasticity of demand for energy is that the magnitude of price-induced changes in demand varies significantly among utility service areas. Nor are price elasticities of demand necessarily constant over a range of price (or income) changes (NARUC, 1977, p. 87).

A computer simulation model of peak and off-peak rates was performed as part of an Ontario Hydro study. The results appear in Table 4.2. This model did not derive elasticities, but rather looked at the relative costs and benefits of time-of-use metering under differing assumptions of rate differentials and elasticities of demand. Even the most cursory comparison of the range of elasticity data between Tables 4.1 and 4.2 will reveal how gingerly projections of cost based on elasticity estimates must be treated.

*See also Appendix 1 to this book.

Long-run elasticity is also influenced by the potential for interfuel substitution—that is, the opportunity to choose appliances with respect to the availability of alternative fuels. The short-run relative inelasticity is influenced by the inability to alter the stock of appliances, so price response must be met by altering intensity of appliance usage or time-of-use alone.

Results of the Vermont experiment indicated that the effects of price elasticity of demand tended to create a deliberate effort to conserve during peak hours and a corresponding incentive to overconsume during the off-peak hours. The Vermont study found a "use as much as you want for as long as you want" attitude prevailing during the off-peak hours. They concluded that "basically, the off peak rate is believed to be responsible for a net energy loss, instead of resulting in energy conservation" (U.S. FEA, 1977b, p. 27).

INTERFUEL SUBSTITUTION

The cross-elasticity of demand for electricity is defined as the ratio of the percentage change in electricity consumption to the percentage change in the price of a competing energy source. Under normal conditions this cross-elasticity is positive, because as the price of alternative fuels rises, people are inclined to consume more electricity.

The decision to substitute fuels is made on the basis of the average price of fuels for those making an initial investment decision, and based on the end-block rates for those contemplating fuel switching. As long as the end rate for electricity is less than the cost of the fuel oil or gas alternative, switching will not occur. One danger is that application of marginal resource pricing in the end blocks could result in customers leaving the system for other fuels with the result that the remaining fixed costs are spread over a smaller number of users. Larger industrial users are the first to leave the system as many factories are equipped to change fuels at short notice. For this reason, the market can be stabilized if the tail block is set at the cost equivalent of the next best alternative source of energy. [This argument is based on testimony concerning natural gas rates and interfuel substitution presented by Reinbergs before the Public Service Commission of Wisconsin in 1979 (Wisconsin PSC, 1979a, d,f). The principle is valid for electricity pricing as well as gas pricing.]

In considering the potential interfuel substitution effects of a new electricity rate, care should be taken to consider its effect on the development of alternative energy technologies, such as solar heating, as well as co-generation in the commercial and industrial sectors.

Carpenter (1978, p. 23) has commented on the relationship between load factors and solar energy. He stated that a winter peaking utility may improve system load factor by *firm* (his emphasis) solar heating installations (i.e., those without standby service); by contrast, a summer peaking utility may expect system load factor deterioration from any solar heating installations. At present, rates discriminate between partial require-

TABLE 4.1.   Overview of Demand Studies

Author/ Analysis	Level of Analysis Type of Data	Dependent Variable	Explanatory Variables[a] P_O P_S Y N W A D t	Other	Functional Form/ Estimation Technique	Price Specification[b]
1. Acton, Mitchell, and Mowill (1976)	Residential electricity demand; monthly data for meter read book areas in Los Angeles County, July 1972–June 1974	Consumption by household	P_O P_S Y N W A D		Linear; total energy equaling the product of the appliance stock and its utilization rate. Does cross-sectional and time-series analysis separately.	Deals explicitly with declining block rate schedule, estimating marginal rate, and fixed charge
2. Anderson (1972)	Residential electricity demand 50 states, 1969	Annual consumption per flexible (new) customer in kWh/customer year	P_O P_S Y W D		Log-linear, OLS	TEB (1000 kWh/mo) TEB (500 kWh/mo) average revenue
	California, annually 1947–1969		P_O P_S Y t			
3. Anderson (1973)	Residential appliance, electricity, gas demand, 50 states, 1960–1970	Shares of appliance stock by energy type for various uses	P_O P_S Y W D		Log-linear, OLS and GLS static and dynamic formulations	TEB (1000 kWh/mo) TEB (40 therms/mo)
		Demand for electricity per household	P_O P_S Y W D			
		Demand for gas	P_O P_S Y W D			
4. Balestra (1967)	Residential/commercial gas demand 49 states and Washington, DC 1950–1962	Gas in Btu × 10^{12}; incremental demand and total demand	P_O Y N W D	Lagged dependent variable	Linear, log-linear, first differences, 2-SLS, maximum likelihood. Partial adjustment formulations	Average price
5. Cargill and Meyer (1971)	Electricity demand for all sectors 2 SMSAs, monthly data, January 1965–December 1968	System load at $t = i$, $i = 1, \ldots 24$, deseasonalized	P_O P_S Y t	Employment of production workers in manufacturing, Y^2	24 hourly equations, OLS	Average revenue aggregated over all classes of customers

Elasticities							
Own-Price		Cross-Price				Stock	Additional
L.R.	S.R.	L.R.	S.R.	Income	Other	Treatment	Remarks
−0.70	−0.35		0.71	0.38(S.R.)		Develops appliance stock estimate based on average monthly consumption; air conditioners and heating weighted by cooling and heating degree days	Elasticity estimates from pooled data. Cross-sectional elasticity estimates for different months show widely time-varying results
−0.91		0.13		1.13	−0.85 size of household.		Separate rows of results for analysis of 50 states and California
−0.88		0.17		0.34 −0.46	0.18 winter temperature. 0.83 summer temperature.		
−1.12 −2.75	Negative but insignificant for oil, electricity, and coal			0.80	−0.94 size of household 0.76 % single family units 0.27 oil 0.12 coal	In share equations	Separate rows of results for the three dependent variables
Elasticities vary considerably by state and groups of states pooled by homogeneity in weather, gas availability, and time period −0.68	−0.03						Incremental demand elasticities significantly greater than total demand elasticities. Substitutes are oil and coal.
	−0.06 to −0.58			Insignificant			Deals explicitly with time-of-day pricing. Expects more price responsiveness if time-of-day pricing existed. Significant price changes should lead to decreased consumption

(continued)

TABLE 4.1. (continued)

Author/ Analysis	Level of Analysis Type of Data	Dependent Variable	Explanatory Variables[a]		Functional Form/ Estimation Technique	Price Specification[b]
			P_O P_S Y N W A D t	Other		
6. Fisher and Kaysen (1962)	Residential electricity and appliance demand 47 states 1946–1957	Demand for electricity in the short run (kWh) given fixed appliance stock	P_O Y A D	Number of wired households, number of marriages.	Multiple regression and covariance analysis (of groups of states). OLS on first differences of the logarithms.	Average revenue
		Demand for appliances in the long run	P_O P_S Y N A	kWh consumed per average use, appliance prices, permanent income		
7. Griffin (1974)	Residential electricity demand, United States' aggregate annual data, 1949 and 1951–1971	Demand per capita	P_O Y A		25 equations, block recursive. Almon lag. OLS and 2-SLS. "Linked" to macro model	Average revenue
8. Halvorsen (1973)	Residential electricity demand 48 contiguous states, annual. 1961–1969. Pooled time series and cross-sectional data	Average consumption of electric energy/customer	P_O P_S Y W A D t	Degree of urbanization, appliance prices.	Static equilibrium model. Loglinear, 2 SLS. IV for static equation using data 1961–1969.	Has separate price equation for marginal price, but uses TEB and average price for demand
9. Houthakker (1951)	Residential electricity demand 42 provincial towns, England, annual data, 1937–1938	Average annual consumption of electricity for customers on two-part tariff.	P_O P_S Y W A		Log-linear	Marginal prices (from two-part tariff), lagged two periods

Elasticities						Stock	Additional
Own-Price		Cross-Price					
L.R.	S.R.	L.R.	S.R.	Income	Other	Treatment	Remarks
	inel. −0.16 to −0.25			0.07–0.33	Appliance prices have no effect. Changes in number of wired households and number of marriages have significant positive effect	Distinguishes short-run demand as a function of intensity of use of present stock vs. long-run demand as a function of those variables that influence rate of growth of stock of appliances	Substantial difference between regions of the country. As regions "mature" economically, price sensitivity decreases. Emphasis that relative and not absolute changes are of importance; hence, most variables expressed in terms of relative changes
inel.		inel.		Current and permanent income important			
−0.52	−0.06			0.06(S.R.) 0.88(L.R.)	0.22 (capital stock)	Air conditioners explicitly included	Only residential electricities reported here. Model intended for simulation. Forecasts to 1981. Study also discusses large users.
−1.0 to −1.21		0.049 to 0.088		0.47–0.54			Deals explicitly with simultaneity problems. For the price equation specified, the use of marginal or average price will yield the same elasticities.
	−0.9 to −1.04		+0.2 to +0.28	1.01–1.17		"Heavy" equipment only as measured in terms of kWh rating	Tentative investigation on seasonality (hours of daylight, average temperature).

(continued)

TABLE 4.1. (continued)

Author/ Analysis	Level of Analysis Type of Data	Dependent Variable	Explanatory Variables[a] P_O P_S Y N W A D t	Other	Functional Form/ Estimation Technique	Price Specification[b]
10. Houth-akker, Verle-ger, and Sheehan (1974)	Residential electricity demand 48 states, 1961–1971	kWh consumption per capita	P_O Y	Lagged dependent variable	(1) Log-linear, partial adjustment model. GLS, error component technique (2) OLS with and without same intercepts (3) IV (lagged Y, lagged P, population)	Differences between TEB for (1) 500 and 100 kWh; (2) 250 and 100 kWh; (3) 500 and 250 kWh as estimates of marginal price
11. Mount, Chapman, and Tyrrell (1973)	Residential Commercial Industrial Electricity demand, 48 contiguous states, annual, 1946–1970	Total electricity demand, kWh $\times 10^6$	P_O P_S Y N W	Lagged dependent variable, appliance prices	Log-linear specification, constant elasticity model—OLS, variable elasticity, model—OLS and IV. Uses error components model	Average price
12. Mount and Chapman (1974)	Residential electricity demand U.S. contiguous states, annual, 1963–1972	Total electricity demand, kWh	P_O P_S Y	Number of customers, appliance prices	Geometric lag, GLS, random cross-sectional effects, log-linear	1. Marginal 2. Average 3. Average/ marginal
13. Taylor, Blatten-berger, Verleger (1977)	Residential electricity, gas, oil demand 50 states and Washington, D.C. 1955–1974	Residential consumption of gas, electricity and oil by state	P_O P_S Y W D	Lagged endogenous variable	Log-linear and linear traditional flow adjustment model; log-linear and linear Koyck model	Marginal and fixed charge electricity rates; average gas price
		Appliance stock utilization rates	P_O P_S Y W A	Lagged endogenous variable	Log-linear and linear flow adjustment and Koyck model	
		Capital stock	P_O P_S Y W	Lagged endogenous variable, appliance prices	Log-linear OLS, error components model used throughout	

| Own-Price | | Cross-Price | | | | Stock | Additional |
L.R.	S.R.	L.R.	S.R.	Income	Other	Treatment	Remarks
−1.0	−0.089			S.R. = 0.143 L.R. = 1.6			Forecasts and backcasts surprisingly good, for so simple a model. Little homogeneity among states. Flow adjustment model, the most useful specification. Elasticities are W.R.T. the different price definitions.
−1.2	−0.094			S.R. = 0.127 L.R. = 1.6			
−0.45	−0.029			S.R. = 0.145 L.R. = 2.2			
−1.2	−0.14 to −0.36	0.2	0.02	0.21(L.R.) 0.02 to 0.10(S.R.)	Unitary (L.R.) W.R.T. population, inel.		The absolute value of the price elasticity positively correlated with price. Elasticities refer to residential, commercial and industrial. IV and OLS give similar L.R. elasticities but different S.R. elasticities.
−1.6		0.05		0.88(L.R.)			
−1.8		0.06		0.65(L.R.)	W.R.T. price of appliance (L.R.)		
−1.7	−0.31	0.03 to 0.61	0.01 to 0.16	S.R. = 0.16 L.R. = 0.61			Discusses effect of increases in P_O on generating capacity and on primary fuel requirements. Also gives equations for commercial and industrial sectors.
−0.82	−0.08			0.10(S.R.) 1.08(L.R.)	Electricity fixed charge −0.02 (S.R.) −0.17 (L.R.)		Reported elasticities for fuel demand are those for electricity
−0.12 to −1.0	−0.06 to −0.54			0.0004 to 0.38 (S.R.)	−0.08 (L.R.) electricity fixed charge	Appliance stock variable a weighted stock using "normal" use as weights	
	−0.02 to −0.22		0.02 to 0.10	Varies widely by appliance; all inelastic however			Varying elasticities for refrigerators, freezers, room air conditioners, water heaters, stoves, automatic washers, dryers, central heat, and central air conditioners

(continued)

TABLE 4.1. (continued)

Author/ Analysis	Level of Analysis/ Type of Data	Dependent Variable	Explanatory Variables[a]								Other	Functional Form/ Estimation Technique	Price Specification[b]
			P_O	P_S	Y	N	W	A	D	t			
14. Wills (1977)	Residential electricity demand; consumption data for 39 Massachusetts electric utility districts and 57 residential rate structures, 1975	Monthly consumption, kWh	P_O	P_S	Y			A	D		Appliance saturation rates used	OLS, log-linear	Marginal and fixed charge electricity rates; average gas price
15. Wilson (1971)	Residential electricity demand 77 cities in 1966	kWh per household and appliance demand	P_O	P_S	Y	N	W		D			Linear, log-linear OLS	TEB for 500 kWh/mo

Source: Hartman, 1979, pp. 444–447. Reproduced, with permission, from the *Annual Review of Energy*, Volume 4. © 1979 by Annual Reviews, Inc.

[a]P_O = own price, P_S = price of substitute fuels, Y = income, N = population, W = weather/temperature, A = stock of appliances, D = demographic/housing characteristics, t = trend.

[b]TEB is typical electric bill (FPC) and TGB is typical gas bill (from BLS).

The following are references for Table 4.1:

Acton, J. P., B. M. Mitchell, and R. S. Mowill (1976), *Residential Demand for Electricity in Los Angeles: An Econometric Study of Disaggregated Data*, Report Number R-1899-NSF, Rand Corporation, Santa Monica, Calif. December.

Anderson, K. P. (1972), *Residential Demand for Electricity: Econometric Estimates for California and the U.S.*, Report Number R-905-NSF, Rand Corporation, Santa Monica, Calif. January.

Anderson, K. P. (1973), *Residential Energy Use: An Econometric Analysis.* Report Number R-1296-NSF, Rand Corporation, Santa Monica, Calif. October.

Balestra, P. (1967), *The Demand for Natural Gas in the U.S.*, North-Holland, Amsterdam.

Cargill, T. F., and R. A. Meyer (1971), "Estimating the demand for electricity by time of day," *Applied Economics*, October, 3:233–246.

Fisher, F. M., and C. Kaysen (1962), *A Study in Econometrics: The Demand for Electricity in the U.S.*, North-Holland, Amsterdam.

Griffin, J. M. (1974), "The effects of higher prices in electricity consumption," *Bell Journal of Economics and Management Science*, Autumn, 5(2):515–539.

Elasticities						Stock	Additional
Own-Price		Cross-Price					
L.R.	S.R.	L.R.	S.R.	Income	Other	Treatment	Remarks
−0.08		0.32(S.R.)				Handles appliance stock through saturation rates in equations.	
−1.33	0.31			−0.46		Appliance stock is a function of "life style." Separate equation, dependent variable: % houses with at least a unit of appliance i. (i = 1 . . . 6) 6 different categories	Price is the major determinant

Halvorsen, R. (1973), *Long-run Residential Demand for Electricity*, Discussion Paper Number 73-6, University of Washington, Institute for Economic Research.

Halvorsen, R. F. (1973), *Short-run Determinants of Residential Electricity Demand*, Discussion Paper Number 73-10, University of Washington, Institute for Economic Research.

Houthakker, H. S. (1951), "Some calculations on electricity consumption in Great Britain," *Journal of the Royal Statistical Society*, Series A, 114, Part III.

Houthakker, H. S., P. K. Verleger, and D. P. Sheehan (1974), "Dynamic demand analyses for gasoline and residential electricity," *American Journal of Agricultural Economics* 56:412–418.

Mount, T., D. Chapman, and T. Tyrrel (1973), *Electricity Demand in the U.S.: An Econometric Analysis*, Report Number ORNL-NSF-49, Oak Ridge National Laboratory, Oak Ridge, Tenn. June.

Mount, T., and D. Chapman (1974), *Electricity Demand Projections and Utility Capital Requirements*, Cornell University Agricultural Economics Staff Paper Number 74-24.

Taylor, L. D., G. Blattenberger, and P. K. Verleger (1977), *The Residential Demand for Electricity*, Report Number EA-235 by Data Resources Incorporated to Electric Power Research Institute, January.

Wills, J. (1977), *Residential Demand for Electricity in Massachusetts*, Working Paper Number MIT-EL 77-016W, MIT Energy Laboratory, Cambridge, Mass. June.

Wilson, J. W. (1971), "Residential demand for electricity," *Quarterly Review of Economics and Business*, Spring, 11(1):7–19.

TABLE 4.2. Summary of Ontario Hydro Study Results

Run	Differential (cents/kWh)	Elasticities			Conservation (GWh)[b]	Net Cost	Additional Cost (millions of 1978 dollars)	Total Saving	Net Saving	Benefit/Cost Ratio
		Peak	Off-Peak	Cross[a]						
Single-rate Metering 1–6						631				
7[d]						1353				
Optional 1	0.7	−0.3	−0.2	0.1	0.008	630	1.31	1.83	0.52	1.4
Two-rate 2	0.7	−0.5	−0.3	0.1	0.014	630	1.31	2.25	0.94	1.7
Metering 3	0.7	−0.5	−0.3	0.3	0.018	629	1.31	2.71	1.40	2.1
4	1.0	−0.3	−0.2	0.1	0.49	562	40.3	109	68.7	2.7
5	1.0	−0.5	−0.3	0.1	0.85	537	40.3	134	93.7	3.3
6	1.0	−0.5	−0.3	0.3	1.09	508	40.3	162	121.7	4.0
7[d]	1.0	−0.3	−0.2	0.1	0.49	1098	37.6	293	255.4	7.8
Full 1	0.7	−0.3	−0.2	0.1	3.16	666	411	376	−35	0.9
Scale 2	0.7	−0.5	−0.3	0.1	5.55	497	411	545	134	1.3
Two-rate 3	0.7	−0.5	−0.3	0.3	7.07	310	411	732	321	1.8
Metering 4	1.0	−0.3	−0.2	0.1	4.51	501	413	542	129	1.3
5	1.0	−0.5	−0.3	0.1	7.93	260	413	784	371	1.9
6	1.0	−0.5	−0.3	0.3	10.11	−7[c]	412	1050	638	2.5
7[d]	1.0	−0.3	−0.2	0.1	4.5	211	368	1510	1142	4.1

Source: Ontario Hydro, 1976h, p. 110.

[a] Peak/off-peak cross elasticity.

[b] Gigawatt-hours = millions of kilowatt-hours (included in saving).

[c] Saving outweighs additional cost plus single-rate metering net cost.

[d] Similar to run 4 but using 5% discount factor.

ments service and full service, and it has been proposed that these requirements be abandoned to encourage solar installations. Carpenter is opposed on the grounds that the cost burden for solar would be shifted to all other customers because of the inclusion of the higher demand charges implied by marginal cost pricing.

CHANGES IN RESIDENTIAL LOAD CHARACTERISTICS

Changes in system and residential load curves over time can be used as a measure of the effectiveness of a time-of-use rate schedule. The resulting shifts can also be used as a basis for further rate modifications leading to even greater cost savings.

The relationship between lowered loss-of-load probability and increased energy consumption during the off-peak period should be stressed. Increased off-peak usage has a minimal effect on system LOLP. If system expansion is based on LOLP as a reliability criterion, even increased baseload power demand should not hasten plant expansion (U.S. FEA, 1976b, p. 37).

Observations on the relative abilities of consumers to shift power seem to indicate that medium-sized customers respond best to time-of-use rates. This implies that "low usage customers probably have fewer appliances with which they can shift their load, and high usage/high income customers appear to have less incentive from a price standpoint to shift their load" (U.S. DOE, 1979?, p. I-5).

Some loads, such as lighting or radio and television, are not deferrable in the sense that customers can use the power involved during an off-peak time. Other loads are much more flexible, such as dishwashing and laundry. Some knowledge, therefore, is necessary concerning appliance stock and income before an accurate assessment of response to time-of-use rates can be made.

A frequently voiced concern involves the incidence of "peak chasing," that is, the phenomenon of new peaks at the periphery of the old peak. The Arkansas study found that new peaks created outside the designated peak period by the time-of-day group were almost as large or larger than the peak demand of the control group during the peak period. Despite this, the experimental group did reduce *both* peak-period energy use and demand, as well as reduce their overall kilowatt-hour consumption (U.S. DOE, 1979?, p. V-1).

The phenomenon of peak chasing is said to occur if the cost of consumption during a narrowly defined peak is significantly higher than the cost of consumption during the immediately preceding time period. This is so because it is easier to shift consumption by, for example, 1 hour than 6 hours (Ontario Hydro, 1976g, p. 9, fn. 1). Peak chasing can be cured to some degree by adjusting the rates or slightly staggering peak hours for different customers.

Green Mountain Power of Vermont also found a sharp increase in demand immediately following the end of the on-peak period. They observed that the differential between contribution to group peak and system peak loads was not seasonally sensitive, indicating that the shifted load is primarily related to hot water heating and other household functions not related to season (U.S. FEA, 1977b, p. 25).

All three classifications of customers in the Vermont experiment utilized *more energy* during the *off-peak rate* period than the previous year: electric heat customers, 12.4% more; hot water customers, 8.8% more; and the total group together, 10.5% more than the previous year. Comparing heating degree days with the previous year indicated that heating should have been slightly less. The net energy loss has two possible explanations. Off-peak customers' load indicates that these customers consistently reached a high evening peak after peak hours and also consumed large amounts of energy prior to the peak hours in the morning. Part of this is believed to be anticipation of higher on-peak heating costs, that is, overheating part of the house to allow for a long carry-through. This explanation of the phenomenon of high evening and early morning consumption is supported by the absence of such practices in summer (U.S. FEA, 1977b, pp. 26–27).

Another result confirmed by both the Arkansas and Connecticut experiments was the permanence of the shift away from the peak period. It was previously believed that customers on time-of-day rates would not reduce their demands on days of extreme weather. This proved not to be the case. The Connecticut report concluded that this constancy of response makes the planning of generation based on expected response more sound (Connecticut PUCA, 1977, p. 120).

The Ohio study conclusions are as follows:

1. In 6 of the 8 months examined, the experimental group consumed significantly less electricity during the peak period than the control group. The reduction in consumption ranged from 21 to 38 percent.
2. TOD rates caused a reduction in usage during the peak period but did not affect total consumption.
3. In 6 out of the 8 months examined, the experimental customers used more electricity during the base period than the control group (U.S. DOE, 1978a).

4.2.2 Commercial and Industrial Customer Response

Effect on Profitability

Commercial and industrial customers are usually the first singled out as candidates for load control because they have the metering equipment in place already, even though industrial customers are least responsible for seasonal peaking. The commercial and industrial customers who are not able to shift their loads face disproportionate increases in their electric bills under time-of-use tariffs (Walters, 1978, p. 18). This may be true, but some amount of energy conservation could potentially offset, partially or wholly, the unfavorable impacts of time-of-use pricing. Industrial customers may be able to reschedule processes, whereas the less flexible commercial customers cannot. It is expected that commercial customers would respond to time-of-day rates "by increasing their thermal insulation levels, decreasing lighting or heating and cooling levels, or by finding a more energy-efficient way to perform a service" (U.S. FEA, 1976b, p. 49). There remains a great lack of knowledge on the part of plant owners of when and where electricity is used and how that use can be controlled (Efford, 1978, p. 62).

In sum, increases in the price of electricity should not seriously affect firms' profitability. Data from the U.S. Census of Manufacturers indicate that electricity costs average about 1% of value added for U.S. manufacturing. Some industries, such as petroleum refining and industrial inorganic chemicals, have greater relative electricity costs (5 and 15%, respectively), but these are also some of the industries observed to make the greatest adjustments to peak-load pricing in Europe (Acton et al., 1978a, p. 20).

The ECAP Study found that a "change to marginal-cost-based rates would have no significant effect on employment or on the profitability of customers' operations in the [Canadian] economy" (O.E.B., 1979, p. 171). Ontario Hydro believes that a shift to marginal cost-based rates would not constitute a major incentive for the relocation of industry, with the exception of *some energy-intensive industries*. The Ontario Energy Board felt that lower rates in the future (no explanation) and the encouragement of energy substitution technology could be offsetting influences in support of industrial growth in the province (O.E.B., 1979, p. 172). However, the board provided scant explanation for these predictions.

The Ontario Hydro Impact Study, which used an interview technique to examine the response to rate increase scenarios by industries (no explicit elasticity of demand data were used), concluded that rate increases alone would cause a shift in the competitive position and reduction in sales of particular industries (synthetic abrasives, for example) (Ontario Hydro, 1976j, p. 28). It found that international and interprovincial companies will gain an advantage over intraprovincial industries (if electricity costs rise faster in Ontario than elsewhere) because of a flexibility in relocating. Some firms, it discovered, had already made provisions for contracting electricity in the State of New York at lower rates than those projected for Ontario (Ontario Hydro, 1976j, p. 28). A related finding was that the higher electricity prices would make electricity self-generation more common.

Other past studies have concentrated on explaining kilowatt-hour sales, not kilowatt sales, and forecasting how much revenue would be lost as prices increased. These studies modeled the electricity economy and derived estimates of demand rather than observed behavior under experimental conditions as was done in the U.S. Department of Energy studies. (A discussion of econometric studies of the demand for energy in the United States and Canada is included in Appendix 1.) The 1976 study conducted by National Economic Research Associates (NERA) for Ontario Hydro concluded that U.S. and Ontario price elasticities of industrial demand for electricity were very similar (although both results were considered to be overestimates of actual elasticities) (Ontario Hydro, 1976d, p. 56).

Of most interest in the NERA econometric model was the attempt to determine the degree of substitution and/or complementarity among factors of production (e.g., capital and labor and energy) in view of increasing energy costs.

"The NERA model projects growth in usage by major industrial users as a function of the growth in output, the growth in the average price of electricity and the growth in the average price of alternative fuels" (Ontario Hydro, 1976d, p. 66). Because the outcome consistently underforecasts growth, it was concluded that elasticities derived from the study were overestimates due to the omission of potentially important variables such as the price of capital, labor, and materials. This and other forecasting models are of value for reasons other than projecting the impact of time-of-use or other electricity pricing schemes. The NERA model concludes that "higher electricity prices will reduce growth in industrial demand . . . within selected manufacturing industries . . . [and] will bring about increased electricity conservation" (Ontario Hydro, 1976d, p. 74).

Two other econometric models worthy of review pertain to industrial demand for electricity. The first is the Berndt and Wood study (1975) entitled "Technology, Prices and the Derived Demand for Electricity." The conclusions of the study are that:

1. Energy demand is responsive to a change in the price of energy with elasticity approximately -0.47.
2. Energy and labor are substitutable with the cross-price elasticity of energy with respect to labor equaling $+0.18$.
3. Energy and capital show a degree of complementarity with cross-price elasticity of energy and capital at -0.18.
4. Capital and labor are found to be substitutable.

These results lead the authors to conclude that because capital and energy inputs are complementary goods, while labor and energy are substitute goods, the lifting of price ceilings on energy types would tend to reduce the energy and capital intensiveness of producing a given level of output and increase the labour intensiveness. Moreover, since investment tax credits and accelerated depreciation allowances reduce the price of capital services, [the complementarity finding] implies that these investment incentives generate an increased demand for capital and for energy. To the extent that energy conservation becomes a conscious policy goal, general investment incentives may become less attractive as fiscal stimulants [Berndt and Wood, quoted in Ontario Hydro, 1976d, pp. 103–104].

A second important study is the Fuss and Waverman report on "The Demand for Energy in Canada" (1975a). While the Berndt and Wood study used U.S. manufacturing industries as a source, the Fuss study used an aggregation of Canadian industries. Since the technology used in the two countries is the same, the results should not be significantly affected by the choice of sample. The Fuss study is directed to two questions: First, what is the effect of a change in the price of a given fuel on the costs of production; and second, what is the effect of a change in the price of fuel on the demand for labor and capital.

Deriving the cost of energy as a function of the cost of competing fuels, Fuss and Waverman estimate an elasticity of demand for electricity of -0.4, which is approximately the same as that resulting from the Berndt and Wood study. Other conclusions differ, however. Fuss and Waverman found that energy and capital were to some extent substitutable, whereas Berndt and Wood found them to be complementary. The difference is important; if the factors are found to be substitutable, continued economic growth is possible even if the price of energy continues to increase. If they are complementary, an increase in the price of energy would lead to a decrease in the demand for both energy and capital, which might slow economic growth (Ontario Hydro, 1976d, pp. 106–107).

The NERA analysts reviewing these studies concluded that these findings can be reconciled because energy is often used with capital to replace labor and can also be conserved by increasing capital investment. Inasmuch as capital can be used to conserve energy, the two are to some extent substitutable.

The analysts caution that estimated price elasticities cannot be considered reliable guides for policy decisions (Ontario Hydro, 1976d, pp. 108–109). This is so because the price elasticity of energy decreases as the share of energy used in production increases. (The high energy cost in the industries sampled may have been due to inexpensive energy prices during the period studied.) The aggregation of industries in both studies disguises the different elasticities among individual types of firms.

The effect of electricity price changes on the industrial demand for energy will yield different results if one is considering price increases of

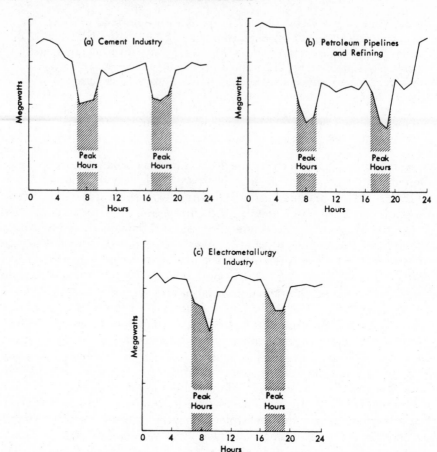

FIGURE 4.4. Winter weekday load curve for selected French industries. (Source: Mitchell and Acton, 1977, p. 44.) © The Rand Corporation. Reprinted with permission.

electricity alone or all energy inputs, or if one is discussing price changes (up or down) resulting from load management techniques.

Commercial and Industrial Demand Response

Higher electricity prices alone can lead to increasing reliance on industrial co-generation. The "addition of cogeneration customers tends to reduce electric utility prices, load factors, oil and gas use and base load capacity additions" (U.S. DOE, 1979d, p. I-8; see also Ontario Hydro, 1976j). Electricity rate reforms such as higher demand charges and demand ratchet provisions, which seek to reduce rates to high-load-factor industrial customers, reduce the incentives for real industrial co-generation.

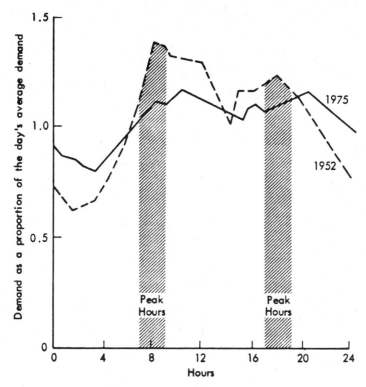

FIGURE 4.5. French daily load curves for representative January workdays, 1952 and 1975. (Source: Mitchell and Acton, 1977, p. 42.) © The Rand Corporation. Reprinted with permission.

Rising relative electricity prices alone will cause fuel switching. This practice takes time for adjustment as capital equipment is usually fuel specific. Generally, industries allow the existing machine to depreciate, then replace it with one utilizing the preferred fuel.

A Rand report concerning nationwide energy and capacity savings possible from the use of peak-load pricing based its conclusions on the high degree of adaptability exhibited by European firms. The analysis was based on data for production processes that did not differ materially between European and North American firms during the mid-1970s. The study assumed that the load curves of U.S. manufacturing industries would assume the load shape of their counterpart European industries with total electricity use remaining the same (Acton et al., 1978a).

The study concluded, "relatively complete adaptation of industrial processes, equipment and work schedules to take advantage of these pricing incentives can be expected within ten years after such rates are first introduced." The pricing approach does not require direct controls on specific end uses of electricity—which imply significant losses in consumer welfare—nor does it require an administrative structure (Acton et al., 1978a, pp. 19–20).

Countries will tend to avoid mandatory requirements that unilaterally increase costs in energy-intensive export industries and that can adversely affect their foreign trade position.

Figure 4.4 illustrates the level of adaptation of major industrial energy users in France (cement, petroleum pipelines and refining, and electrometallurgy) during a winter month when the 7 A.M. to 9 A.M. and 5 P.M. to 7 P.M. peak prices are in effect. Daily load curves for 1952 and 1975 are contrasted in Figure 4.5.

4.2.3 Summary and Conclusions

The exact response to time-of-use rates cannot be determined without experimental work being performed by individual utilities. Although studies report similar trends in customer response, these utility-specific studies have used rate structures based on their own costs, load characteristics, and service population and the resulting conclusions are not easily generalized. A comprehensive information program appears mandatory for any successful plan. A carefully constructed tariff can overcome problems of fuel switching and peak chasing. Rates should also be evaluated for their potential impact on the development of solar energy and industrial co-generation. Imposition of time-of-use rates should not adversely affect firms' profitability and should favorably affect employment (as labor is substituted for energy). If time-of-use pricing is phased in for industrial customers, complete adjustment should be possible by the end of the decade.

5

ELECTRICITY PRICING: EXPERIENCE AND GENERAL CONCLUSIONS

5.1 REVIEW OF INTERNATIONAL PRACTICES

The notable feature of the Japanese power industry is that it runs with a much lower reserve capacity than North American or European systems. Marginal costs were seen to be rising in 1974 because of the high-quality fuel required by enforcement of industrial waste laws and pollution control laws. Transmission and distribution costs were also perceived to be rising as sales increased and power plants were more remotely situated.

The Japanese have abandoned the theory of decreasing marginal costs and have adopted an inverted block tariff system. Prices in blocks with the largest consumption levels are charged prices that reflect (increasing) marginal costs; surplus earnings yielded from excess marginal costs over average costs are allocated to provide a discount on blocks with low levels of consumption. This structure is similar to a lifeline rate that aids minimum power consumers (up to 120 kWh per month).

Japanese per capita power consumption is lower than that of European countries. In Japan, power consumption is highly correlated with the income level of consumers, and large residential consumers are more income elastic but less price elastic in their demand for power than low-income consumers. Therefore, because Block II (121–200 kWh) and Block III (over 201 kWh) differ little in price, and Block III is only 10.5–16.7% above Block II, there is little impact of the inverted electric tariff system on overall power demand. Japanese utilities have found that industrial customers are less price elastic to power demand and that their ratio of power cost to total production cost is generally low. Therefore, industrial consumers were able practically to disregard the potential impact of the inverted electricity rates except for high-energy-consuming industries such as aluminum, where inverted rates may damage competitiveness (Osawa, 1978, p. 96).

One reason for the low reserve capacity in Japan (i.e., 8–10% versus 15% U.S. average and 45% for Ontario Hydro) is the use of pumped storage operations for hydro power. Other factors contributing to this almost complete utilization of capacity include (1) wide area coordinated system operations ("electric power companies are expected to deal effectively with contingencies and take any actions which may facilitate inter-regional cooperation") and (2) load regulation (special contracts to large load manageable consumers, plus seasonal- and time-differentiated rates). In general, inverted rates were chosen because they best encouraged energy conservation and contributed to the welfare of the lower-income class. Implementation of inverted rates has not been as effective in curbing electric power demand because its price elasticity is small compared with income elasticity and production elasticity (Osawa, 1978).

Electric utilities in Japan and France offer "demand-limiting" tariffs to their residential customers. Residential customers subscribe to a specified level of demand that they may not exceed. A demand-limiting tariff allows the utility to assess demand charges without installing expensive metering devices.

The Tokyo Electric Power Company, the largest utility in Japan, has used a two-part residential tariff since 1976. "Customers are assessed a fixed monthly charge based on their contract demand plus a charge per kilowatt-hour for all kilowatt-hours they use. The fixed monthly charge is directly proportional to the maximum subscribed demand. Energy charges are based on an increasing block schedule" (U.S. DOE, 1979c, p. III-3) and are the same for all residential customers, regardless of the level of subscribed demand.

The French residential tariff replaces the idea of a subscribed capacity charge (which requires a meter capable of measuring demand in each time period) with a fuse or circuit-breaker charge, which increases with the maximum current rating regardless of time of use. The utility supplies these master fuses and thus controls each resident's maximum demand. There is a small initial block that captures fixed costs, and then a flat rate per kilowatt-hour. An optional time-of-day use is available at about half the standard rate (peak from 6 A.M. to 10 P.M.) and has a higher initial block to cover the meter costs. About 18% of all customers (2.7 million) take advantage of this rate, primarily those with storage heaters and water heaters (Mitchell and Acton, 1977, p. 34).

Norwegian residential tariffs are based on high winter system peaks similar to those prevailing in Canada. The subscribed maximum demand charge reflects the system peak. In addition, there is a per kilowatt-hour energy charge as well as an additional charge for demand in excess of the subscribed load. An optional ammeter is installed in the consumer's kitchen to indicate when demand is exceeding the subscribed rate and when excess charges are being tabulated (Mitchell and Acton, 1977, p. 22).

Swedish residential tariffs also use a subscribed demand charge scaled to the size of the main fuse amperage. A flat rate per kilowatt-hour is levied for energy. Optional time-of-day rates are available for residences with storage heating. This rate includes a fixed charge to recover metering costs. A variety of other tariffs is also available from which households can choose, depending on their load factor (Mitchell and Acton, 1977, p. 24).

In both France and Sweden, the provisions for five-year contracts for levels of maximum demand serve to avoid abrupt changes in revenue and load shape (Mitchell and Acton, 1977, p. 46).

British residential tariffs consist of a simple flat rate per kilowatt-hour. An optional rate is available under which all night usage is billed at slightly more than half the per kilowatt-hour of peak-period usage. The costs of the dual register meter for this type of rate are charged to customers. Storage heating (radiators) has been promoted actively since 1961, and as of 1977, 14% of all residential customers in England and Wales had some form of storage heating (Mitchell and Acton, 1977, p. 39). (The effect of storage heating on load curves was illustrated in Figure 4.3.)

Conclusions

European experience with innovative tariff structures is longstanding and offers several choices of models. Predominantly hydro systems like those in Norway and Sweden are similar to their Canadian counterparts. The use of voluntary programs and optional rate structures affords a gradual and cautious means of exploring rate reforms at little expense.

5.2 SELECTED U.S. ELECTRIC UTILITY RATE DEMONSTRATION PROGRAMS: SUMMARY OF FINDINGS AND CONCLUSIONS

The Electric Utility Rate Demonstration Program was undertaken to remedy the lack of empirical data available to utilities and regulatory agencies on the effects of innovative pricing strategies. The proposals studied included various load management techniques as well as time-of-use rates. General conclusions derived from the studies are that:

1. Customers respond significantly to changes in electricity price by time-of-day.
2. Time-of-use rates reduce residential customer peak demands even during the most weather-sensitive peak periods.
3. Customer attitudes toward time-of-use rates are largely positive.

A brief description follows of the structure and conclusions of some of these studies.

ARIZONA

The Arizona experiment was a very modest attempt to determine the effects of time-of-use rates. Twenty-eight rates were tested, with only five families sampled per rate. The applicability of the data obtained is highly limited by the extremely small sample size and the fact that a substantial number (68%) of those in the presample were rejected for various reasons.

The final conclusions were that the participating customers reduced their electricity consumption during the peak period by 7–16% and by 1–9% during the intermediate period. Overall electricity consumption declined slightly. No statistically significant price elasticities were observed for time-of-use prices, a finding that was largely dismissed by the shortcomings of the experimental design. Contributing to the tenuousness of the conclusions was the experimental practice of calculating rates so that the average customer would save 30% over his previous baseline bills. The study further provided that no bills would be any larger than the baseline period, no matter how much additional electricity was consumed. This offered little incentive for reduced consumption.

The lesson to be learned from this particular experiment is that a study must be extremely well designed to yield any valid results. Federal funding of this study totaled $210,000, and its conclusions are applicable only to a small subpopulation of the cities of Phoenix and Yuma, Arizona.

ARKANSAS

The Arkansas study tested time-of-use rates by day and season as well as increased tail block tariffs. These rates were derived from the utilities' long-run incremental costs. The study spanned 28 months. Customer participation was mandatory, and no personal interviews were conducted (so information on appliance stock was not obtained).

Study findings and conclusions from the final report follow (Arkansas PSC, 1979?, p. I-5):

Consumers are willing to shift their electrical usage patterns when given pricing alternatives as incentives.

The primary source of short-run price elasticities appears to be in air conditioning, as demonstrated by the load shifts during the summer months and system peak day.

Medium-size customers appear to have the highest price elasticities of demand. This appears reasonable, since low-usage customers probably have fewer appliances with which they can shift their load and high-usage/high-income customers appear to have less incentive from a price standpoint to shift their load.

The percentage reduction in daily peak demands was approximately the same for peak days as for the average summer day. This differs from previously held beliefs that customers on time-of-day rates would not reduce their demands on the hottest days of the summer.

The fact that the peak demand of the residential time-of-day (TOD) customers occurred outside the peak period, yet is as high as the control group peak demand in the peak period, leads us to believe that the peak/off-peak ratio used in the experiment may have been too high.

CONNECTICUT

The Connecticut experiment tested a three-part peak-load pricing rate with a 16:1 peak to off-peak differential and a monthly charge of $2. The customer sample was divided into five consumption strata, depending on prior years' electricity consumption. Fifty customers in each of the strata were chosen at random, with 199 ultimately participating and 184 finishing the experiment. The test lasted for one year (October 1975 to October 1976) and cost approximately $1 million. The test did not address issues of price elasticity of response to different price levels but sought instead to examine the load characteristics of residential customers purchasing electricity under a time-of-use rate.

Detailed conclusions and supporting statistics of this extensive study follow (our emphasis added) (Connecticut PUCA, 1977, pp. S-6–S-9):

1. *Distribution of consumption by rate periods:* The portion of the test group's total electricity usage in the peak hours was 10 percent in contrast to 13 percent for the control group, representing a difference in the percentage of consumption used on peak of 23.1 percent. In the intermediate period the test group consumed 47.6 percent of its total usage and the control group 49.6 percent, indicating a difference in the percentage of consumption used in those hours of 4 percent. In the off-peak period the percentage of usage by the test group was notably higher than that of the control group, 42.4 percent in comparison to 37.4 percent, or a difference of 13.4

percent in the distribution. *This evidence and the responses to the customer survey confirm that the test rate induced customers to shift the use of appliances out of the peak hours and to a lesser extent out of the intermediate hours into the off-peak hours.*

2. *Load shapes:* Profiles of the hourly average level of demand of the test group throughout the day, called load shapes, were notably different from those of the control group and the test group in the prior year and confirmed the decided shift away from the system peak periods. *The test group peak for an average January day occurred in the evening in the hours ending at 10 and 11 P.M., in contrast to the control group whose peak occurred in the hour ending at 7 P.M., nearly coincident with the time of the system evening peak* in the hour ending at 6 P.M. In response to the lower prices in effect until 8 A.M. the test group displayed early morning loads in the winter which were considerably higher than those of the control group. In the summer, the test group reduced its load during the system peak hours, 10 A.M. to 12 noon and 1 to 3 P.M., in contrast to the control group whose load remained relatively constant through those hours.

3. *Load coincident with system peak:* The test group loads coincident with the January evening and the June system peaks were substantially lower than those of the control group—1.559 kW vs 1.871 kW in January, 0.654 kW vs 1.106 kW in June. The test group loads were also lower in the hour coincident with the system peak in each month of the test.

4. *Load factor:* The annual load factor of the test group based on its load coincident with the January evening system peak was 0.597 in contrast to 0.524 for the control group, or a 14 percent improvement. *In all months of the year the test group load factor based on its load coincident with the system peak of the month was higher than that of the control group.* The difference was greater in the warm months than in the cold months.

5. *Test group peaks:* The monthly peaks of the test group were lower than those of the control group in all highly weather-sensitive months as were the winter and summer peaks.

6. *Diversity:* There was no conclusive evidence whether the test rate had the effect of encouraging a greater coincident use of appliances among the test group customer and thus reducing the diversity of appliance use found with non-time-differentiated rate structures. In four of the seven most weather-sensitive months of the year there were improvements in diversity as reflected in lower coincident factors; in the other three months, diversity decreased, as indicated by the higher coincidence factors for those months.

7. *Reduction in use:* The average test group customer consumed 5 percent less electricity than did the average control group customer. Because of statistical differences between the test group

sample and the control group sample it is not clear what portion of the difference is directly attributable to the rate. Thus, *there can be no firm conclusion regarding the effect of the test rate on conservation in the sense of reduction of use.*

8. *Customer response on days of extreme temperature:* A concern of generation system planners in connection with peak load pricing is whether the customer response will be consistent on days of extreme temperatures. If customers were to use a larger portion of their total daily consumption in the peak hours on those days than on other days (if, for example, they "paid the price" and kept the air conditioner going more than usual in peak hours on the hottest days) the result would be a sharply higher peak on those days than on average cold or warm days, reaching a level not anticipated if system expansion plans had been made on the assumption of a consistent customer response to a peak load pricing rate. The proportions of total daily electricity consumption occurring in the four peak hours on the days of the January evening system peak and the June system peak (which are days of extreme temperatures) were nearly equal to the corresponding proportions on average January and June weekdays respectively, indicating that *the test customers responded similarly to the pricing "signals" on extremely hot and cold days as on days of normal seasonal weather.*

9. *Differences in response among strata:* All the strata were influenced by the test rate, but the response differed in type and degree. The major differences in the response to the test rate of small users (Strata 1 and 2)—generally those using electricity for a refrigerator, television and miscellaneous small appliances—and large users (Strata 3, 4 and 5)—those who usually had electric water heating and ranges and space heating (the latter mostly in Stratum 5) and more intensive use of other appliances—were as follows:

   a. *Distribution of consumption by rate periods:* Customers in test group strata 1 through 4 all shifted a similar proportion of consumption away from the peak periods and used from 10.0 to 10.6 percent of their total consumption in peak hours in contrast to 12.5 to 13.3 percent of consumption in the peak hours by the equivalent control group strata. In addition, the percentage of consumption by these strata of the test group during the intermediate hours was less than that of the control group, but the shift was not as pronounced as for the peak hours. *The largest users, Stratum 5 customers, were able to reduce their share of consumption in the peak hours to a greater extent than other strata,* and only 8.8 percent of their consumption occurred on peak in comparison to 12.5 percent of the consumption of Stratum 5 control group customers. *Stratum 5 of the test group also consumed a substantially larger share of electric energy in the off-peak hours than did any of the other strata* (46.2 percent in contrast to from 40.8 to

43.1 for the other strata) and also consumed more off-peak than Stratum 5 in the control group (46.2 percent versus 40.5 percent).

b. *Load factor based on load coincident with system peak:* The load factors of each of the strata based on their loads coincident with the January evening and the June system peaks were significantly higher than those of the corresponding control group strata. In relation to the January evening system peak the load factor of the first three strata improved more decidedly than those of Strata 4 and 5. By contrast, in June the improvement was greater in the higher use strata and less in Strata 1 and 2.

c. *Diversity:* As with the test group as a whole, no conclusive pattern emerges. However, the results suggest improvement in diversity for the smaller users and deterioration for the larger users.

d. *Reduction in consumption:* Strata 2 and 3 of the test group consumed less electric energy during the test year than did the equivalent strata in the control group. The larger users in the test group, Stratum 5, used more than those in the corresponding stratum of the control group. Stratum 4 consumption was about the same for both groups. Sampling constraints precluded valid comparison of the consumption levels in Stratum 1.

10. *Customer attitudes:* A detailed survey of test customers' attitudes following the test revealed that a *large majority responded favorably to the test rate and considered that the test allowed them to make savings in their electric bills* compared to what they thought the bills would have been under the regular rate. A large majority was willing to continue in the test even with the possibility of a rate change. In general, *the response to a time-of-day rate schedule was strongly favorable.*

11. *Schedule changes:* The customer survey revealed that test customers responded to the test rate structure by changes in household schedules rather than by changes in activities. Nearly all test households reported making changes in their household schedules or patterns of living: 36 percent of the test households said they made "many" changes, 59 percent reported making "some changes." Only 5 percent of the customers interviewed said that they made no changes. Sixty-two percent of those that made schedule changes said they shifted the time they did their laundry; 31 percent changed the time they cooked meals, 29 percent changed the time for cleaning up and using the dishwasher, 12 percent changed the time they ate meals or watched television. The test sample generally found it easier to control the use of the air conditioner than that of the electric heating system to respond to the rate.

12. *Use of time-control or storage equipment:* Very few test customers installed time-control devices to facilitate adjustment to the rate. None invested in heat or cold storage equipment which could have maximized the benefit to be gained from the low off-peak prices.

13. *Short-term energy shifts:* Because of the short duration of the test, the test provides no conclusive evidence regarding customer shifts to appliances using other forms of energy (e.g., from electric to gas ranges) or to electric appliances that can be operated advantageously primarily off-peak (e.g., time-controlled hot water heaters).

14. *Revenues: A comparison of the revenues yielded by the test rate showed that the average test customer did in fact pay less on the test rate than would have been paid on the regular rate.* The test rate did not yield revenues equivalent to those yielded by the regular rate. This can be explained by the fact that the experimental rate was designed in the absence of information about customer response and only reflected an untested estimate of the distribution of use by customers within the three time periods.

The results of the test answered the major questions regarding customer acceptance and response. It can be concluded, therefore, that residential customers of all levels of use are able to respond to a rate structure based upon time periods with different prices. The majority of the test customers appeared to welcome the opportunity to make an economic choice concerning the time they used electricity.

OHIO

In the Ohio experiment, 100 residential customers were placed on a seasonal, cost-based, time-of-use rate with a peak/off-peak rating period in effect during weekdays. The study was directed at evaluating customer acceptance and ability to alter usage patterns. The experiment lasted for 18 months, was voluntary, and was carried out without customer compensation. The results, as are all conclusions cited herein, are applicable only to the sampled population involved (U.S. DOE, 1979e, pp. 4–6).

1. *Peak Period.* During the six-hour period of high prices, customers on the TOD rate used significantly less electricity than customers on the existing (declining block) rate. The reduction in usage was relatively consistent over the time period examined (which encompassed all seasons). *Monthly reduction in usage during the peak period ranged from 21 to 38 percent* [our emphasis], or from 1.52 to 2.62 kWh per customer per day.

2. *Base Period.* Some shifting of usage from the peak to the base period occurred, as evidenced by an increase in usage during the base period in seven of the eight months examined. . . . Statistically, these increases were not significant. However, such small shifts in

usage are not detectable with the present sample sizes and the large variation in usage from customer to customer observed during the base period. . . .

3. *Usage During the Hour Preceding/Following the Peak Period.* During the one-hour periods preceding and following the peak period, customers on the TOU rate schedule did not use significantly more electricity than the customers on the existing rate schedule.

5.3 FINAL CONCLUSIONS AND IMPLICATIONS

Some change in the current structure of electricity rates is inevitable. Marginal cost pricing in a practical form (i.e., time-of-use pricing) is advanced as a means of partially reconciling high and rising fuel and building costs with consumers' desires for lower fuel bills. It is able to reconcile high costs with low rates by reducing customers' demands on generation systems when they are most expensive to operate. Under assumptions of a level load, generation costs can be cut and capacity additions can be postponed or even cancelled.

Because it is not always known whether capacity expansion is due to increased peak consumption or to overall growth in electricity demand, a wise electricity pricing policy would seek to reduce both peak-period demand and overall demand. An increased price is the most effective means of achieving a reduction yet the most unacceptable from the consumers' point of view.

This dilemma can be overcome by offering an element of choice so that high rates are not perceived as unfair or punitive. Several forms of rates have been in effect abroad that are worth testing in North America. Some of these rates can be implemented easily and with little risk to the utility. Seasonally differentiated time-of-use rates that require no additional metering can reflect more accurately a utility's cost of service while offering consumers a choice concerning their electricity consumption. Ratchet rates reflecting the previous year's billing data and perhaps a discount/surcharge applied relative to changes in consumption are another accessible method of altering consumers' behavior at little expense to the utility. Voluntary programs should be tried with a variety of rate structures to test their feasibility.

Relative to the political and environmental costs of additional power facilities, the benefits of conservation and load management programs are extremely attractive. The price of electricity generation is higher than formerly realized, and this fact must be conveyed to purchasers.

The growth of electricity demand will be accompanied by the growth of nuclear and/or coal-fired power plants. Indirectly, therefore, the policy taken on electricity pricing determines the rate at which these two generation options expand.

6

RESIDENTIAL ENERGY
CONSERVATION:
A CALIFORNIA CASE STUDY

INTRODUCTION

Until very recently, California led all other states, as well as the U.S. federal government, in the scope of its energy conservation legislation. It will be apparent, however, that many of the critical program proposals remain in the recommendation category or have suffered defeat at the hands of legislators or the public. Notable among these are proposals to bring the benefits of energy conservation to California's renters (estimated at 40% of the population) or to make appropriate insulation a precondition of sale or transfer of homes. The imagination and strength of the California Energy Commission's (CEC) policy recommendations, coupled with the tight energy situation within the state, should ensure enactment of vigorous conservation initiatives in the future. The policy instruments with which these initiatives are pursued—either incentive or mandate—will be a central focus of this chapter.

6.1   THE CALIFORNIA ENERGY SITUATION

6.1.1   The Growth in Demand

It was not shortage per se but the alarming growth in energy demand that initially led the California Assembly to commission a study of electricity demand. That study, *California's Electrical Quandary* (Rand, 1972, p. iii), made it clear that the energy policy of the state should not be limited to accommodating the demand for electricity (Council of State Governments, 1976, p. 4). Previously, total energy use had increased by 14% annually from 1968 to 1973 and per capita electricity use had grown at an average annual rate of 3.6%. In 1974 statewide electricity demand was projected to increase at an annual rate of 5.7% over the next 20 years (CEC, 1981, p. 2). In fact, the 1979 Biennial Report found residential average energy consumption declining because of higher energy prices, the saturation of energy-using appliances, low population growth, and adoption of conservation regulations (CEC, 1979a, p. 1). The latest 20-year forecast of annual electricity sales growth is 1.44% (CEC, 1981, p. 2). There exist, however, substantial differences between growth projections made by the CEC and the utilities, although by October 1980, California utilities had cut their estimates of the levels of demand in the early 1990s by more than half (CEC, 1981, p. 2). An overestimate of only 0.5% could result in the unnecessary addition of two large power plants at a cost of $2–4 billion (CEC, 1979g, p. 2).

California is particularly dependent on oil and gas. Together they provide over 93% of the state's total energy needs and 72% of the energy used to generate electricity (CEC, 1979a, p. 19). The natural gas situation has improved considerably because of supply agreements with Mexico and Canada and an overall decline in consumption. But supply projections are extremely tentative, ranging from a low of 2.431 billion cubic feet to a high of 6.573 billion cubic feet by 2000 (CEC, 1981, p. 115). The picture is made more uncertain because of the pricing provisions contained in the Natural Gas Policy Act, which base gas price increases on the relative price of fuel oil.

The petroleum situation is less encouraging. Total petroleum consumption has doubled since the early 1960s and in 1978 was a billion barrels. Transportation is the largest and fastest-growing energy end-use sector. By the year 2000 the number of drivers and automobiles in California is expected to increase by one-third (U.S. Dept. of Transportation, 1980, p. 14). Because responsibility for transportation lies outside the legislative mandate of the CEC, it has been less active in this area. By and large, it has been the U.S. federal government that has attempted to offset the growth of gasoline sales by the mandatory 55 mph speed limit, fuel efficiency standards, and the "Gas Guzzler Tax."

6.1.2   The Role of Energy Conservation

Because one of the major tasks of the CEC is to forecast energy demand, it has built into its models projections of demand that consider its own energy conservation programs. Whether these econometric projections constitute valid evaluations is open to question, but even the least optimistic estimates of nonprice conservation measures alone project a savings of 10–12% of electric sales by the year 2000 (CEC, 1979g, p. 66).

The "high conservation" scenario is based on the assumptions that maximum household appliance efficiency standards are in effect, that 90% of new multifamily units have solar water heating, that ceiling insulation retrofit programs reach 100% of the dwellings, and that CEC air conditioning cycling (load management) programs affect 8% of the state's air conditioners. Under these assumptions, the 1978–1984 growth rate would be reduced by 45%. This high conservation scenario, applied to the commercial and industrial sector (assuming maximum effectiveness of utility energy audit and appliance and building standards) would lead to a 20% savings in electricity and natural gas use over the same period (CEC, 1979g, p. 92).

The commission concludes that conservation measures have the most impact on future electricity sales. If recent conservation standards (1979) were to be completely ignored, the 1978–2000 growth rate would be almost 40% higher. Table 6.1 summarizes the scenario impacts on electricity consumption. Clearly, the level of conservation efforts is the most critical factor affecting demand. The commission cautions, however, that under the assumptions of the high conservation scenario and leveling energy prices by the mid-1980s, consumers generally might become lax again in their energy conservation habits. How deeply these habits are ingrained or institutionalized will ultimately test the validity of today's efforts.

6.1.3   The Energy Market in California

The major criterion for judging the effectiveness of energy conservation measures is that they be cost effective. The term itself is open to interpretation, particularly with regard to the mandate of the California Energy Commission. The CEC has the authority to make regulations, but only insofar as they are "cost effective" to the consumer. In fact, cost effectiveness at the macro level has not been a sufficient condition for commission action; proposed regulations must be demonstrated to be cost effective and of immediate benefit to consumers at the household level.

The issue of cost effectiveness brings into consideration the relative prices of energy sources against which conservation expenditures are

TABLE 6.1.   Scenario Impacts on Electricity Consumption in California

Scenarios	Statewide Annual Growth Rate in Electricity 1978–2000 (%)
Baseline	2.00
High oil price	1.90
Low oil price	2.19
High economic growth	2.25
Low economic growth	1.75
High conservation	1.37
Low conservation	2.75

Source: CEC, 1979g, p. 93.

measured. Current mispricing of all energy—electricity, gasoline, and natural gas—distorts the price signals received by consumers and makes conservation options less attractive than they would otherwise be.

In the electric power sector, there are a number of pricing mechanisms that are inconsistent with conservation efforts. Some, like declining block rates and average cost pricing in the face of rising marginal costs, are familiar policy issues. Others, such as various forms of income tax subsidies, are less publicized. A report to the CEC concerning the comparative economics of solar energy, Alaskan natural gas, and nuclear power found that the utilities and oil companies enjoy income tax subsidies to such an extent that no utility pays taxes on revenue from new investment in gas or electricity. This reduces the nominal cost of Alaskan gas by approximately 30% and the cost of nuclear power by 25%. The prices paid for new gas and electricity are reduced an additional 27% and 25%, respectively, by pricing tax subsidies through which users of existing inexpensive utility energy pay higher rates to reduce the rates charged to new users (CEC, 1979d, p. 5).

Additional subsidies include corporate capital tax subsidies, accelerated depreciation, investment tax credits, and interest deductions. In California the tax life of a nuclear power plant is 16 years. All these subsidies must be added to the market price of energy to determine a "true economic cost." Even this cost does not include pollution and other environmental effects. Subsidies to energy conservation could be viewed, therefore, as a counterbalance to the subsidies enjoyed by energy producers.

In fact, the CEC views itself in a brokerage role, particularly in overcoming market imperfections in the introduction of new energy technologies, including solar. Considered on a full-cost basis, given California's 55% tax credit on solar installations, solar power proved 7% less expensive than Alaskan gas and 89% less expensive than nuclear power. But these cost comparisons remain hidden or are unimportant to the consumer.

The public does have rising energy prices on which to base its decisions, however inaccurately those prices reflect actual cost. But the public must be convinced that the real price of energy will rise much faster in the future before undertaking conservation measures. If energy prices are perceived as part of a general inflationary trend, cost/price signals again lose some of their impact.

Two examples illustrate different consumer responses to inflation that hamper the adoption of new energy sources and energy conservation behavior. The first involves evaluating solar systems, which are highly capital intensive, at the nominal rate of interest when costing on a deflated basis would enhance solar cost competitiveness. The second example involves cases in which the consumer is aware that increases in the cost of energy, in real terms, are consistent with other price increases. For example, the cost of gasoline has trebled since 1960 in California, but in real terms, the cost increase was only 2% over the same period (CEC, 1979k, p. 1–7).

6.2   THE DIVISION OF RESPONSIBILITY FOR ENERGY CONSERVATION WITHIN THE STATE

6.2.1   The California Energy Commission

The California Energy Commission came into existence as a result of the Warren–Alquist State Energy Resources Conservation and Development Act in 1974, a comprehensive energy bill passed in the aftermath of the Arab oil embargo. Utilities interested in streamlining siting procedures for new generation facilities, conservationists concerned with protecting the California coastline, and antinuclear groups opposed to the siting of additional nuclear plants formed the unlikely coalition that supported the bill.

The legislation, as well as the primary jurisdiction of the commission, involved an extensive electrical energy management program: planning and forecasting, energy resource conservation, power plant siting, and research and development.

The commission is presided over by a five-member team that, it was reasoned, would be able to withstand political pressure. Because the members are appointed by the governor, the commissioners tend to reflect the current governor's political biases. The CEC is funded by a $0.0001 per kilowatt-hour surcharge on electric bills throughout the state and its budget is supplemented by federal funds.

Two features unique to the conduct of the commission's business are the requirements for an elaborate biennial report incorporating both a forecast of energy demand and an evaluation of its programs and for an office of Public Advisor who ensures that all relevant public interest groups are advised of upcoming hearings and decisions. California functions under an "open government" convention, where all decisions must be made in public sessions for which at least seven days' notice must be given.

Essentially, the Warren–Alquist Act made the provision of an adequate supply of electrical power the responsibility of the state, so the Commission's primary regulatory responsibilities lie in the electricity sector.

The commission's first biennial report in 1977 formulated its energy conservation strategy in terms of improved efficiency and a long-term public education program. The purpose of the information-gathering and forecasting responsibilities was to derive independent estimates of energy consumption and to provide a data base against which to monitor and evaluate conservation programs. Although the commission is not committed to any one means of promoting energy conservation, its origins as a regulatory agency tend to focus its approach on standards setting for appliances and building design characteristics and on directing utilities' efforts to improve the efficiency characteristics of the existing stock of homes and appliances.

6.2.2   The Public Utilities Commission

All investor-owned utilities in California fall under the jurisdiction of the Public Utilities Commission (PUC). This accounts for about 90% of all utilities. The remainder are municipal utilities that are self-governing.

Both the Energy Commission and the PUC share jurisdiction over the utilities. Viewed from this perspective, it is easy to understand why the majority of conservation efforts are performed by individual utilities at the direction of the EC and PUC. The combined authority of these two commissions over utility rates and new generating facilities enables them to require the implementation of energy conservation programs rather than run such programs themselves. By law the utilities were originally forced to conduct conservation programs with "vigor, imagination and effectiveness." This subjective test became an issue in rate requests. As a consequence, the test criterion language was subsequently changed, requiring the utilities to pursue conservation programs to their "fully cost-effective level."

Nevertheless, those government officials with whom we spoke in California concurred that the utilities are doing their utmost to promote energy conservation. The utilities themselves are under severe capital and legal restraints. The National Energy Act prohibits the use of oil and gas in new generation facilities, and the widespread use of coal is prevented by California's own air quality standards and lack of resources. In effect, energy conservation represents expanded capacity to the utilities or even an opportunity to diversify into what the CEC envisages as "energy service organizations." The potential scope of utility activities includes the following:

1. Consumer information (conservation tips, using past utility bills for comparison, etc.).
2. Conservation services (adjustment of pilot lights and thermostat setting).
3. Direct marketing (ceiling insulation, water heating insulation, water flow restricters, setback, etc.).
4. Financing conservation measures.
5. Utility coordination in programs and power sharing.

The fairness and wisdom of utilities' participation in the conservation market, and solar market in particular, will be discussed later. This was the subject of two extensive investigations by the PUC. The utilities are required to submit annual reports evaluating all conservation efforts as well as ten- and twenty-year estimates of potential energy savings from those programs.

6.2.3 CALTRANS and the California Transportation Commission

The California Transportation Commission was established in 1978 to unify California transportation policy. It acts in an advisory capacity to the legislature, as opposed to CALTRANS (the California Department of Transportation), which is actually responsible for the conduct of transportation activities within the state. Because the Energy Commission does not have full authority over the transportation sector, its activities have been limited to funding energy conservation programs within CALTRANS and evolving a full-fledged transportation division. Actual and contemplated transportation conservation activities within the state are described in Section 6.6.

6.2.4 The Federal Government

California's energy conservation plans interact with federal plans in several different ways. In some instances the two programs complement each other well, such as the federal R&D expenditures on alternative energy sources and the California emphasis on commercialization of those technologies. By contrast, other policies are in direct conflict, such as California emissions laws, which lower mileage efficiency. In general, the federal program emphasizes economic incentives whereas the state places priority on information, education, and technical assistance.

Where the state conservation plan promotes efficiency in the use of electrical energy, the National Energy Plan emphasizes reduced levels of use per se, that is, curtailment of use of petroleum and natural gas.

6.3   INCENTIVES AND OTHER POLICY OPTIONS

Because the Energy Commission is in a position to command the resources of California's utilities, with few exceptions it has chosen to use nonfinancial incentive measures in the conduct of its own programs. This tendency is accentuated by California laws requiring state constitutional amendments when public funds are used for private purposes, such as loans, grants, or rebates.

Two defeated pieces of conservation legislation involved the sale of $25 million in bonds to finance insulation and solar information programs and a loan provision to finance installation of solar heating. Although these ostensibly would involve no cost to the public, the CEC did not have the public support to carry the measures.

The commission has lost other battles in its attempts to shape energy conservation legislation. One cause to which the commission has been devoted since its inception is mandatory insulation before the sale of a home can be completed. This was defeated three times by a real estate lobby because it was seen as complicating the transfer of property.

The regulatory route is no easier than the incentives route, as both require political muscle. Incentive measures that have failed during the course of the past few years include the following:

AB 774:
(1975)   Permits the EC to fund the construction of prototype solar housing, among other measures.

AB 1131:
(1975)   Appropriates $10 million from the general fund for solar and insulation demonstration projects.

AB 1464:
(1975)   Permits a deduction from taxable income for the total cost of purchase and installation of solar systems.

ACA 30:
(1975)   Allows exemption from property tax of property using alternative energy sources for energy generation.

SB 216:
(1975)   Exempts from taxation property used as a solar energy system (e.g., greenhouses and solaria).

AB 3044:
(1976)   Allows a tax deduction for the purchase and installation of solar or wind devices.

AB 2982:
(1978)   Sets up a Warranty Insurance Pool to cover warranties of defunct solar companies by a surcharge on collectors.

AB 3255:   Allows utilities to establish test market programs for sell-
(1978)   ing solar equipment and charging the costs to ratepayers.

Most often the commission uses a combination of efficiency standards, consumer information, and backup R&D to further its program goals rather than direct incentives or subsidies. The notable exceptions are the 55% tax credit for solar heating and the 40% tax credit on residential and nonresidential conservation investment. Although the commission is wholly committed to solar power, it advocates the use of passive, or architectural, design features. This approach lends itself more easily to establishing building standards for new construction; the utilities then would presumably evolve retrofit programs individually. No efficiency standards have been developed for industrial processes or the transportation sector, although both are under study.

Information and technical assistance programs may be used to complement incentives and regulations or may be adequate in themselves to stimulate conservation actions that are clearly cost effective. The commission undertook a major attitude survey in the belief that its information programs should be directed toward a specific class of consumer whose energy-consuming behavior is well understood. The study found no correlation between family size and total energy consumption; rather, it found that energy use increased with the age of respondents. There was no clear-cut relationship between affluence and conservation, but prior research identified conservation behavior with consumers who had knowledge of conservation practices. Studies cited on the relative effectiveness of monetary or other incentives found conservation behavior to increase if the incentives were given soon after conservation takes place (CEC, 1978c, p. 5).

Respondents were also asked to identify their preferred group to solve current energy problems. The utilities and private industry ranked one and two above government, which was seen as ineffective. Strong preferences were found regarding load management alternatives (in particular, appliance cycling programs*), and, of the mandates considered, the public found the requirement that new homes and appliances be energy efficient most acceptable. Solar systems were thought too expensive to purchase and install, as well as being unattractive, and were considered more controversial and less effective than insulation measures.

All CEC initiatives share the following characteristics: a preference for demonstration of currently feasible technology; local community orientation; utility roles under public sector guidance and oversight; and the setting of efficiency standards. Because of the commission's inability to finance major programs, it has concentrated its efforts on low-cost

---

*This term applies to the utility-controlled remote cycling of appliances as part of a load management program.

strategies that can produce benefits in the near term. Many of these strategies involve removing barriers to alternative energy development and aiding commercialization of existing technologies.

In addition to the aforementioned CEC initiatives, the commission offers a number of indirect residential conservation incentives, including helping small energy-related firms qualify for small business loans, sponsoring a state-wide passive solar design competition (Assembly Bill 3046), and offering technical assistance to architects, engineers, and local building officials in building design and revision of local building codes.

The commission additionally conducts tests on the performance of various energy conservation materials and equipment and, through the Testing and Inspection Program for Solar Equipment (Assembly Bill 1512), provides consumers with reliable comparative data on solar collectors. Energy conservation standards for new buildings are of two alternative types: "prescriptive," where the individual components of the building must meet certain specifications, or "performance," where building design is computer analyzed for efficiency.

6.4   CALIFORNIA SOLAR PROGRAMS

6.4.1   Goals and Constraints

Assembly Bill 3324 required the commission to develop a plan for the maximum feasible implementation of solar energy in California. The commission had planned to increase its development program in the 1978–1979 fiscal year by increasing staff from 70 to 106 and by increasing contract dollars from \$2.08 to \$3.26 million, but these proposals have temporarily been abandoned in the wake of Proposition 13. In terms of technical potential, "solar-assisted systems using known technology could provide at least 70–75 percent of each residence's water heating requirements and at least 50 percent of space conditioning requirements," where feasible (CEC, 1977a, p. 73). It is also recognized that the use of solar energy is potentially four times as labor intensive as electrical energy (CEC, 1979d, p. 37). This is of particular interest to governments concerned with the broad social issues associated with alternative methods of energy provision.

Today, California leads the nation in the use of solar energy, with approximately one-fourth of the U.S. installations (CEC, 1981, p. 25). The present goal is 1.5 million solar installations by 1985, which represents 20% of the residential and commercial buildings. This goal can be achieved only if solar is economically preferable to gas or electric heating and if prices will lead consumers to adopt the least costly methods. Absent this cost differential, the state will be forced to rely on a compensating financial incentive. These incentives must be designed with sufficient flexibility to be changed periodically in response to evaluation findings, changes in energy prices, or preemptive federal initiatives. The California solar tax credit ends December 31, 1983, and the federal solar tax credit (30% of the first $2000 and 20% thereafter, up to $10,000) will be phased out by 1984, when the solar industry is presumed to be competitive. With the various utility and government subsidies, a homeowner may have out-of-pocket expenses of as little as $265 for a typical $2500 active solar system. Under the California PUC solar financing program, some homeowners can receive an additional $20 per month credit on their gas utility bills for four years (CEC, 1981, p. 156).

A joint inquiry by the PUC and CEC into the development of solar power in California stressed that a higher priority should be given to large residential complexes and commercial buildings than to individual homes as a means of accelerating the development of solar energy (California, 1978).

The primary constraints to widespread adoption of solar devices include inadequate information for consumers, loan financing difficulty, energy prices that do not reflect relative costs, legal barriers (e.g., "sun rights"), "spillover costs" (i.e., those who would have to pay for conservation do not obtain full energy cost reduction), and uninformed public attitudes. The tax credit addresses the problem of relative energy prices; other policy recommendations described are aimed at the other constraints.

6.4.2   The California 55% Solar Tax Credit

California provides a credit of 55%, not to exceed $3000, for all installations of solar energy devices (residential and nonresidential) on a per-system basis. The tax credit applies to water heating (domestic water heating, swimming pool heating, or pool covers), space conditioning, solar electricity generation systems, wind energy systems, and solar industrial applications. Unlike the federal tax credit, which applies to owner-occupied buildings only, the California credit is allowed to the person owning the property at the time the system is installed. The tax credit in excess of state taxes due may be carried over into subsequent tax years.

Passive solar applications, within defined limits, are eligible for the California credit, although the federal guidelines are unclear on this issue. Swimming pool heaters and covers are not eligible for the federal credit.

In addition to the credit on the solar devices, credit may also be claimed on conservation measures taken to enhance the effectiveness of the solar system, including weather stripping, insulation, and water flow restrictors. Under the federal provisions, a separate credit is allowed for conservation measures and they need bear no relation to solar power.

In addition to the solar tax credit, California utilities have undertaken a variety of solar energy programs.

6.4.3 Evaluation of the Effectiveness of the Solar Tax Credit

By 1981 more than 70,000 solar installations, a large proportion of which were swimming pool covers and heaters, had received the solar tax credit. The 55% credit was an amendment to an initial bill that would have provided only a 10% credit.* A study commissioned by the CEC after the 10% credit was implemented found that at least a 25% credit would be necessary to encourage solar installations.

The 55% credit was enacted in October 1977 and made retroactive to January 1977. An analysis of 1977 tax returns, completed in August 1979, revealed the following (CEC, 1979j, pp. 1–2).

Thirteen thousand taxpayers claimed the credit in 1977, two and one-half times more than claimed the 10% tax credit in 1976.

Total credits of $10.5 million were claimed, plus an additional $2.7 million to be paid as carryover.

Of the total number of credits, 99.7% were claimed by individuals; the rest were claimed by corporate taxpayers who installed systems on commercial or corporate-owned single-family residences.

Swimming pool use constituted 72% of the total (of which 42% was for active pool heating systems and 25% was for pool covers).

Domestic water heating systems comprised 14.7%.

Conservation measures were employed by 6.5%.

Four hundred solariums were claimed.

Average system cost (claimed by individuals) was $1472.

Of the credits claimed, 55% were by taxpayers in the middle-income range ($20,000–$40,000), and 31% of the total net credit amount was awarded to taxpayers with income over $50,000 (average family income = $16,000).

There were no photovoltaic or solar roof pond installations.

*Senate Bill 218 allowed a 10% credit retroactive to January 1, 1976. The credit was claimed by 5434 individuals for a total of $435,953.

These findings very quickly branded the tax credit as a subsidy to the rich. Some explanation for the number of pool covers and heaters purchased relates to the fact that California was in the second year of a severe drought and pool owners were forced to cover pools to conserve water, an action that naturally might lead them to consider solar heating. Lifeline rates in effect in California additionally render solar pool heating cost effective as rate payers are reducing their consumption to levels considered essential.

The finding that the median income of individual taxpayers claiming the credit was $30,581 supports the proposition that the diffusion of solar energy is likely to be from the high income groups down and, additionally, that it is still in its "luxury" phase. This finding is also consistent with the fact that homeowners' income is probably greater than that of the population at large, and people in the lowest tax brackets do not pay taxes and, therefore, could not benefit from an incentive in the form of a tax credit.

Although the commission heavily publicized the measure by distributing a half million solar tax credit brochures, it is estimated that about 30% of builders have been unaware of the state solar income tax credit, and builders and homeowners alike do not know how to identify qualifying solar systems (CEC, 1979a, p. 199). Further, when these initial California data were assembled, the federal tax credit had not yet been passed, so no publicity or interaction effects could be evaluated. (The amount of federal solar credits claimed is deducted from the California credit.)

6.4.4   Planned and Recommended Actions

The preferred emphasis of the commission is on "passive systems," although it will continue to test and demonstrate solar technologies, disseminate information, and reduce legal and institutional barriers. This direction is consistent with a number of local ordinances that mandate solar energy features in new homes. Ten local governments now require either solar water heating in new homes, solar swimming pool heaters in new pools, or passive solar design in new construction (CEC, 1981, p. 29).

In October 1975, the city of Davis, California, adopted the nation's first energy-conserving building code, which has served as a model for other efforts throughout the country. The action was prompted by a finding that city residents used a third more electricity per capita than the U.S. average because of poor orientation and adaptation of Davis homes to the local climate. The levels of energy consumption since the institution of the code display declines that are truly dramatic and give further credence to the regulatory aproach. (See Figure 6.1.)

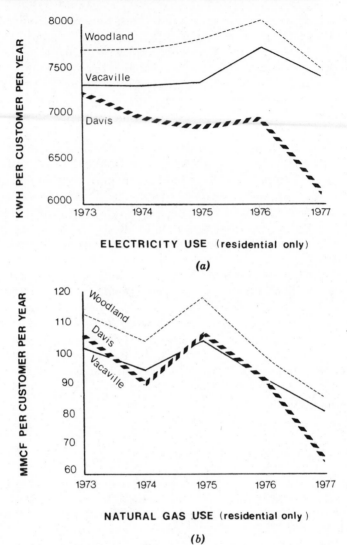

FIGURE 6.1. Changes in energy consumption in Davis, California. (Source: Hunt and Bainbridge, 1978, p. 23; Courtesy of Solar Age Magazine.)

Future directions for the commission's activities include the following:

Establish a local government revolving loan fund to purchase solar retrofits for low-income housing and local government buildings.

Direct the PUC to require utilities to provide subsidies for solar energy.

Establish a solar voucher system where utilities or manufacturers redeem coupons from the state for each system installed.

Provide a negative income tax provision or rebates for solar purchases by low-income households.

Allow a portion of the cost of leased solar hardware to be deducted from a renter's income.

Exempt solar systems from property tax (in fact, the commission found that assessors tend to ignore solar installations in assessments).

Allow a tax deduction based on the savings in conventional fuel costs provided by solar energy.

Allow the tax credit to be used by municipal utilities that lease solar systems, or allow the credit to be passed through to customers.

In the 1980s, efforts in the residential sector will shift from new construction to the retrofit market—developing mass-marketable, passive modules (passive solar bay window kits, attached solar greenhouse kits, etc.) and encouraging the industry rather than the homeowner to shift to solar.

6.4.5   The Utilities and Solar Energy

Initially, the utilities were reluctant to become involved with solar power. A number of consumer groups, afraid of a possible utility monopoly on solar power, eventually brought the issue to the attention of the legislators and the Public Utilities Commission as utilities' involvement increased. One 1978 bill signed by the governor (AB 2984) requires that before a utility begins a solar manufacturing or marketing program, the PUC must find that the proposed program does not restrict the competition or growth of the solar industry. A second bill (AB 2987), which died in the Assembly, proposed that utilities' involvement in the solar industry be outlawed entirely.

The Public Utilities Commission undertook two distinct hearings on the subject—one to determine whether utility involvement in solar energy constituted a possible antitrust infringement and the other to establish the permissible limits to utility financing of solar power.

These hearings were initiated by a state legislature directive to evaluate solar financing options from a market as well as an economic perspective. The Public Utilities Commission was asked to determine the "extent to which each alternative would facilitate the implementation of cost-effective solar energy systems." In conjunction with the PUC, the California Energy Commission undertook a market study during September and October of 1979 in which 2683 people were surveyed and a number of public hearings were conducted concerning attitudes toward solar energy and preferred means of financing.

Some of the preliminary results obtained from the survey indicated that public attitudes concerning solar energy had been shaped predominantly by the media and very little by utilities or state and federal agencies. Respondents felt generally that it was an important energy source (78%), although less than half (42%) felt that it was a viable alternative at that time.

Four financing options were offered to test consumers' reactions. The option drawing the most significant positive response was the "pay on sale of home" option. The other alternatives were a tax credit, low interest rate, and extended term financing. Of those respondents who would definitely buy a solar water heater if a given financing option were available, 38% preferred the "pay on sale of home"; the tax credit and low interest rate options were chosen by 28% and 21% of the respondents, respectively.

Of a variety of "nonfinancing" benefits that would most likely attract consumers to the purchase of a solar water heater, lower utility rates and utility-provided maintenance were the most desired benefits (42% and 29%, respectively); the remaining choices—"cash rebate," "free inspection," "representative visits home"—were chosen as a preferred benefit by 10% of respondents in each case.

An interesting result of the survey was its description of those demographic variables most likely to be found among persons interested in purchasing a solar system. The important variables were:

Age 44 and under.
A household with a full-time employed female.
A household of three or more.
A household with younger children.
A household with college-educated adults.
Incomes over $25,000.
A person or family planning to move soon.

Most people (63%) said they desired more information before they would consider purchasing a solar system. Information on cost was desired by 39% of the respondents; on utility bill savings, by 22%; on efficiency, by 20%; on maintenance, by 20%; and on durability, by 19%.

Of the probable sellers of solar systems, respondents preferred to purchase from a utility company by a slight margin over a solar contractor (40% to 36%). Ten percent preferred to buy a system from a large department store, such as Sears, and 8% said they preferred a heating/air conditioning contractor.

The conclusions from the market survey suggest that a financing program, without more information as to savings and quality assurance, would have some effect on stimulating the market, but only to a limited degree. A more effective approach would be to combine various financing options with information programs that would serve to alleviate consumers' doubts and answer their questions.

The California Energy Commission also tested a large-scale solar financing program via a computer simulation model. The hypothetical market penetration rates studied were 20% in 10 years and 80% in 10 years. Three models were tested:

1. *A rate-based model.* Under the assumptions of this model, the utility rate-bases 50% of the system cost and the remainder of the cost is paid by either the solar consumer or the utility, either of whom could claim the applicable solar tax credits. Installed cost for a single-family dwelling was assumed to be $2500. The original investment of the utility is recouped from the residential solar user at time of sale and credited against the utility's rate base.

2. *A tie-in utility program with conventional lending institutions.* This model is basically a loan program utilizing capital from conventional lending institutions, with the utility subsidizing interest payments. In practice, the program could be administered by either the utility or financial institution.

3. *Direct (rate-based) utility low-interest financing.* This model assumes that the utility obtains capital from whatever source and then pays for the solar system. The amount of the purchase including administrative costs is placed in the rate base. The consumer either owns the system or leases it from the utility. Monthly or annual payments may or may not be required, as well as a loan balance payment on sale of the home.

Whether a financing option is rate-based or not defines whether the expenditure is to be treated as an investment or an expense. If a utility expenditure is treated as an investment, it is placed in the rate base. The sum of investments is the base on which the utility earns its returns. If the expenditure is treated as an expense, it is charged to the rate payers as incurred. The utility earns no return on expenses and therefore the rate payers incur no continuing obligation on the expense. In both cases the utility receives funds through rates to cover the expense or investment. For example, if the utility purchases a $2500 solar system for a customer and treats it as an investment on which it is able to earn a 10% rate of return, the utility may earn 10% per year on the depreciated value of the asset from the rate payers. If it is treated as an expense, the entire $2500 amount will be recovered from rate payers in the year it is incurred without further obligation on the part of the rate payers. Generally, the costs of rate-basing an expenditure exceed the cost of expensing it. This is not only because of the continuing return on the investment charged to rate payers, but also because the utility is taxed on its returns to investment, but expenses are tax deductible.

Of the three financing options, the greatest savings were found in utility tie-in programs in which the utility subsidizes the interest payments on bank financing. "Savings" were measured in kilowatt-hours per year and translated into dollars via the oil equivalent. Long-run capacity savings, which were substantial for the 80% penetration assumptions, were also projected. Because the data were fitted to the utilities' own in-house supply planning computer programs, estimates of costs and savings were realistic for the different utility service areas involved. In each case where a simulation was run (for Pacific Gas and Electric, Southern California Edison, Southern California Gas Co., and San Diego Gas and Electric), net costs of the program were translated to average residential bill increases per month. Under the 80% penetration assumption, electric bills would be raised an average of 8 cents per month in the PG&E area, 4 cents per month in the Edison area, and 19 cents per month in the SDG&E area. Rate increases for gas-using customers would be higher: In the 80% penetration case, residential gas bills would rise by $1.25 per month in the PG&E area, $1.52 in the Southern California Gas area, and $2.23 in the SDG&E area. Rate increases under the 20% penetration assumptions were approximately one fourth of the preceding increases.

Bank financing schemes were assumed to be less expensive because the cost of funds to a bank is lower than that available to utilities; banks typically are financed with a greater proportion of low-cost debt than utilities.

The interest rate chosen for the simulations was 8% (subsidized) and 13.5% (loan interest rate). The lower the interest rate offered to the customer, the higher the cost of the subsidy that must be borne by the taxpayers or rate payers. A 13.5% loan on a $2500 solar water heater subsidized to 6% would represent a $2973 subsidy over the 20-year term; a 13.5% loan subsidized to 8% would represent a $2240 subsidy.

Capacity and fuel savings from a full-scale solar water heater financing program under assumptions of 20% market penetration would equal 322 million kWh per year and 320 million therms per year for the four major investor-owned utilities in the state. This is equivalent to 4.34 million barrels of oil per year; 80% penetration would save the equivalent of 21.4 million barrels of oil per year, or approximately 20% of the oil burned by investor-owned utilities in the state in 1979. Despite the initial rate increases, absolute dollar savings would accrue to rate payers under either of the three financing assumptions. The Energy Commission added that the rate additions from implementation of a full-scale solar development and financing program would increase monthly bills by less than increases sustained during 1979 alone.

Despite initial reservations that utility participation in solar financing would be anticompetitive, no objections were raised by banks on this issue during the hearings. Instead, evidence was submitted that suggested that small loans in the amounts required for installation of a solar system were only marginally profitable for banks that displayed little interest in this potential market.

Other concerns involving the interaction of solar power and power companies relate to the cost of back-up systems. Rates for back-up systems should not be set higher than other residential electric rates or they would constitute a financial disincentive to the installation of solar power. Similarly, necessary quantities of gas and electricity should be supplied to users at prevailing rates. Should solar power develop as anticipated by the CEC, utility service representatives must be trained to identify problems and make minor repairs to solar-assisted space and water heating systems. Finally, master metering, a practice usually condemned by energy conservationists, may be a more effective monitoring technique when solar equipment is added to existing buildings than conversion to individual submetering without solar energy. (The PUC currently requires individual meters but, in cases involving solar installations, the applicant is encouraged to apply for an exemption.)

6.5   LOAD MANAGEMENT PROGRAMS

"If the demand for electricity were level at all hours of the year, . . . California would only need about 23,000 MW of power plants. Instead, [there are] . . . 39,000 MW in large part due to the variability in demand" (CEC, 1979a, p. 183). "In 1972 California utilities projected that peak demand would grow at an annual rate of 7 percent and would exceed 47,000 megawatts by 1980. Actual recorded peak demand in 1980 was 35,000 MW, roughly 25 percent less than the utilities projected" (CEC, 1981, p. 131). This decline was due in part to an average 55% increase in electricity prices experienced by California consumers in 1980.

The CEC was legislatively mandated by Section 24503.5 of the Public Resources Code to develop load management standards by July 1, 1978. Those standards have, in part, been derived from a variety of experimental studies at a cost of one-half billion dollars.

The CEC participated in a variety of different studies under the title "The Electric Utilities Demonstration Project," funded by the federal government and conducted by selected California utilities. Three of the projects have yielded results that have been adopted as required utility practice. They are (1) appliance cycling (offering customers voluntary utility-controlled air conditioner and electric water heater load cycling programs); (2) marginal cost rates (submitting marginal cost rates in all rate proceedings using an approved methodology); and (3) swimming pool pump information and assistance (restricting the use of swimming pool filter pumps to off-peak hours by utility or customer-installed time clocks).

Throughout this study it must be borne in mind that because of the climate, swimming pools and air conditioning constitute tremendous energy demands within the state and are therefore the target of most residential conservation efforts.

6.5.1 Analysis of the Residential Air Conditioner Cycling Program

In 1976 the largest California utilities initiated four different air conditioner cycling experiments designed to measure load impacts and customer response to strategies and types of equipment. In general, the data obtained were relatively consistent with the findings of utilities outside California and with engineering expectations. The results indicated that load reductions in both average peak period demand and maximum customer peak period demand can be obtained from air conditioner cycling programs.

A significant new conclusion reported from the program and not reported in other studies was that cycling reduces total energy consumption as well as peak load consumption. Part of the study consisted of an attitude survey that tested the willingness of participants to continue in the program for a second year. This attitude survey established important relationships between cycle time, load reduction, and consumer acceptance. Table 6.2 summarizes these relationships.

This table shows that load reductions and customer acceptance are dependent on the cycle-off time. Additional data demonstrate that an average of 17% of the customers were willing to have their air conditioners cycled off for the entire peak period. From the 40% cycle-off time and above, acceptance drops off sharply then levels off at the 17% rate acceptance. The correlation between the participation rate and cycle-off time was −0.93.

The question of rebound effect was also studied. This occurs when cycling stops and air conditioner load increases (rebounds) beyond what it would have been without cycling. Cycling into the evening hours suppresses this effect.

There still remain questions concerning cycling that the studies were not able to answer. These involved the dollar value of load-shedding potential to the utility and its overall cost effectiveness, considering the cost of cycling equipment, as well as the value to customers of a regular, reduced level of service, versus total loss of service during a supply outage. When asked if the program should be offered to all qualified

TABLE 6.2.   Relationship Between Air Conditioner Cycle-Off Time, Load Reduction, and Consumer Acceptance

Utility	Percent of Cycle-Off Time[a]	Diversified Load Reduction[b] (%)	Customer Participation in Second Year (%)
SMUD	15.5	not available	99
PG&E	23.3	5–10	95
PG&E	43.3	12–24	82
SCE	50.0	24.32	59

Source: CEC, 1979c, p. 42.

[a]Percent of cycle-off time: the percent of time the air conditioner was switched off by the utility during each half-hour period.

[b]Diversified load reduction: The amount of air conditioner load that was reduced, in excess of natural interhousehold diversity, weighted to reflect a cycling strategy that would prevent a rebound effect; this figure is the reduction that would be experienced during an operational program.

customers on an optional basis, 98% responded positively; only 59% felt the program should be mandatory.

Nominal financial incentives were initially offered to induce participation in the study. On this issue the report (CEC, 1979c, p. 94) concluded that financial reward provides a customer with a payment for achieving a desired action or effort; it is an effective motivator for short periods, but impractical over longer periods or as a means to sustain efforts. Financial incentives were not found to be a primary motivator for initial participation. By 1982 this program, which uses only voluntary participation, was one of the largest in the nation, involving over 90,000 air conditioning systems (CEC, 1981, p. 146).

6.5.2   Load Management Tariff Standard

In the first year of a two-year study conducted by the Energy Commission on marginal cost pricing of electricity, an evaluation was made of all the commonly used marginal costing methodologies with the cooperation of the six largest utilities and six outside consumer groups. The object was to devise a costing method that the commission economists found valid and that would serve as a future basis for marginal cost pricing within the state. The major conclusion was that none of the methodologies could quantify costs adequately for use as a tariff based on marginal costs. The second year of the study sought to derive an alternative method of costing and means of rate design to be used throughout the state.

The study commission never issued a final report; rather, it was instrumental in the Public Utilities Commission's decision requiring California utilities to file marginal cost information at the time that rate applications are made. Thus the PUC adopted the principle of marginal cost pricing as recommended by the Energy Commission. Although specific rates remain unaffected, the marginal cost information requirement was adopted.

Generally, lifeline rates are currently in use in California. The first lifeline rates were enacted as a discount to low-income senior citizens in Los Angeles in 1975. The special rate gave a 50% discount on the first 360 kWh of bimonthly consumption amounting to an average saving of $10; consumption below 360 kWh brings smaller benefits.

In the strictest sense, lifeline rates are promotional rates. They represent the right to purchase a quantity of energy at below cost. The problem with these sorts of residential rate structures, as with the traditional declining-block rates, is that they give the customer no warning that additional growth in use brings into the supply system new sources of gas and electricity that are presumably more expensive than present sources. In California most lifeline rates were imposed on the old declining-block structures so the net effect is that large users actually subsidized smaller users. The "universal lifeline" rate used in Los Angeles (in addition to the previously described rate for the elderly and disabled) was based on a small flat rate, with no initial charge, and does not reflect the increasing marginal costs of new power.

Although the commission supports marginal cost pricing in spirit, using it as a rate basis is another issue. Some critics argue that full marginal cost pricing would be a windfall to utilities and the excess would have to be taxed away and redistributed.

One other utility practice in California that has yet to be remedied involves granting line extension credits (i.e., free line extension footage) to builders, developers, and homeowners in exchange for installation of electricity-consuming equipment such as electric water heaters and clothes dryers.

6.6 TRANSPORTATION PROGRAMS

Transportation is the fastest-growing energy sector in the state. While residential consumption of energy decreased at 3.5% per annum since 1975, energy use for transportation rose over 6.5% per year. Per capita annual gasoline consumption has increased from 340 gallons in 1960 to 533 gallons in 1977, a 60% increase (CEC, 1979a, p. 14). Unfortunately, the Energy Commission, or any other California agency, has yet to devise a scheme for halting the growth in gasoline demand. Of the two following programs, the ride-sharing program and the potential use of registration fees and gasoline taxes to influence consumption, only the former is in effect.

6.6.1 California Ride-Sharing Program

The incentives offered by the CEC and CALTRANS are administrative and organizational; the financial incentives received by the participants stem from their reduced transportation costs and are set forth in Table 6.3. For example, a vanpool of 12 persons reduces automobile travel about 100,000 miles per year, removes approximately 7 automobiles from the roads, and saves each occupant an estimated $1000 per year. Large employers and state and federal offices are the best prospective ride-sharing sponsors. The vans may be either leased (third-party ownership), privately owned, or employer owned. There exists a federal tax incentive program that offers a high investment tax credit to employers who purchase vans.

The program evaluation claimed that 11,500 persons started ride sharing as a direct result of the program and nearly 80,000 applications were processed from April 1977 to June 1978. Although the response and the associated benefits are heartening, they do little to alleviate the real dilemma of gasoline consumption in the state.

6.6.2 Registration Fees and Gasoline Taxes

Senate Bill 269 (1977) directed the California Department of Motor Vehicles (charged with registration and licensing) to develop and submit to the legislature a proposed vehicle registration fee and license fee schedule scaled according to weight, fuel efficiency, or some other criterion. The bill died with no action taken.

The Energy Commission has consistently recommended that vehicle registration be based on fuel efficiency, with rebates given for the purchase of vehicles exceeding federal mileage standards. It has also advocated removing or reducing sales taxes on new automobiles with fuel economy exceeding that required by federal standards.

The interaction between air pollution emissions and fuel economy works against California, and the Energy Commission has advocated that air quality standards be reduced to accommodate fuel-efficient vehicles, or, more politely, "cause vehicle emission control strategies to be developed which allow for the sale in California of the most efficient new vehicles possible." Existing federal law prevents states from setting stricter mileage standards for new automobiles than exist at the federal level. California requested that the new federal "Gas Guzzler Tax" schedule of taxes and rebates be adjusted so that the same incentive and disincentive applies to a given model despite the drop in mileage due to its vehicle emission standards.

The California Transportation Commission, which is a policy advisory body in the area, refused to recommend that vehicle registration vary either with individual air emissions or fuel efficiency. It did state that

TABLE 6.3. Program Benefits: California Ride-Sharing Program, April 1977 Through June 1978

	Dial-in/Mail-in Matching	Organizational Outreach	Vanpools	Buspools	Statewide Totals
Vehicle miles of travel reduced (millions)	30.8	45.5	7.1	7.9	91.2
Fuel conserved (millions of gallons)	1.85	2.74	0.43	0.48	5.50
Pollutants reduced (tons)					
Carbon monoxide	940	1,385	215	240	2,780
Oxides of nitrogen	115	170	30	30	200
Hydrocarbons	150	225	35	40	450
Parking needs reduced	2,600	4,100	600	200	7,500
Direct user savings (thousands)	$3,441	$5,080	$795	$884	$10,200

Source: California Department of Transportation (1979, p. 20).

transportation in California was responsible for 90% of the carbon monoxide, 70% of oxides of nitrogen, and 50% of the hydrocarbons—the three major pollutants affecting air quality. In light of this pollution problem, a case exists for varying registration fees with emissions rather than fuel efficiency, despite the fact that reduced levels of emission also entail reduced fuel efficiency.

There will undoubtedly be a rise in the state fuel tax, but primarily to offset loss of real revenues due to inflation rather than as a measure to discourage consumption. The Transportation Commission posed the question as "Should fuel tax rates be adjusted to offset the effects of inflation on the state transportation improvement program?" (CTC, 1979, p. iii). The state's 7 cents per gallon fuel tax is a specific tax, that is, it is a fixed amount per gallon and, as such, has represented a declining proportion of the price of gasoline over time. It was set at 2 cents per gallon in 1923, rose to 7 cents in 1963, and has remained unchanged since that time. The commission recommended that it be indexed to offset the effects of inflation, which have increased the cost of highway construction in the state by 195% from 1968–1978.

There is also a 6% state sales tax applied to gasoline. The Energy Commission has no stand on the gas tax and so far the legislature has taken no action on indexing the tax for inflation.

The Transportation Commission did, however, recommend that the scope of transportation programs currently funded by 25% of fuel tax revenues be expanded to include all forms of public transit and ride sharing, "but only when transportation revenues are adequate to meet street and highway needs." At present, these revenues may not be used for maintenance and operations of mass transit systems, passenger vehicles, facilities, equipment or services, although they may be used for research, planning, construction, and related administration of mass transit. Basically, CALTRANS' job is the building and maintenance of highways, so it is unlikely that sufficient funds will find their way to mass transit under the current proposal.

Some peripheral programs and recommendations involve endorsement of the use of mileage meters to be included as standard equipment on all new automobiles by 1985. The Transportation Element of the CEC also proposes to increase the availability of "retro" fuel-saving devices for automobiles via certification and direct financial incentives. The Transportation Element additionally assists municipal governments in improving traffic flow on surface roads and has completed a study recommending sodium vapor street lighting to save energy.

The transportation staff of the Energy Commission concedes that the 55 mph speed limit is potentially the largest single source of energy savings in terms of automobile use, but California is the only state that does not use radar to enforce highway speed limits.

6.7 INSULATION RETROFIT PROGRAMS

The National Energy Conservation Policy Act of 1978 requires that utilities offer comprehensive energy audit and retrofit programs to their residential customers. All of the major utility companies in California now have information or marketing programs for residential insulation, but these programs vary widely in approach. The utility companies are free to design their own programs, the programs are all voluntary as far as the customer is concerned, and the only real effectiveness criterion employed by the utilities is that the cost of the programs be less than the cost of new generating facilities. In exchange for vigorous conservation programs, the utilities are allowed a 0.05% upward adjustment of their allowable rate of return.

PG&E is the largest utility in the state, with over 3 million electric customers and 2.7 million gas customers, and is judged to have the best insulation program in the state. Its program initially offers a free home energy audit, suggesting how energy savings could be gained through various conservation measures and recommending an optimum level of insulation. If the customer is interested in purchasing insulation, PG&E offers the names of two prescreened contractors from a revolving list who are obligated to submit estimates when asked. These contractors are screened both as to the quality of their work and their price competitiveness; they charge in the lowest one-half of all contractors.

After insulation is installed, PG&E conducts a thorough inspection and the contractor is required to correct any shortcomings detected. Initially, there was a high failure rate because of poor workmanship and materials. A good inspection program is the key ingredient to a successful retrofit operation not only to ensure conservation of energy, but also to prevent potential firehazards from improperly treated materials or interference with electrical wiring.

The most urgent need is for a program that reaches renters. The commission considers that the failure of the market to reward renters and landlords for using energy efficiently has created a need for government intervention—either by regulation or incentive. However, the commission concludes, "Nevertheless, any legislation requiring energy conservation retrofit should permit landlords to pass the fair cost of conservation measures along to the renter, at a rate commensurate with the renters' savings from reduced utility bills" (CEC, 1979h, p. 1–30). This strategy could easily distort the intentions of such legislation if renters were asked to bear the price of retrofit in an amount equal to their utility bill savings with no net incentive to conserve.

Under the first PG&E plan, if homeowners wished to finance the cost of insulation, the utility charged them 8.5% interest with 60 months to pay. The amount was added to their regular bill and service could not be interrupted for failure to pay this amount. The default rate was negligible.

PG&E now offers a zero interest financing program, known as ZIP, for customers installing a package of residential conservation features such as insulation and weatherstripping. The ZIP program is administered by a wholly-owned subsidiary of PG&E and, by the end of 1982, paid out over $44 million in loans.

The commission has relentlessly advocated mandatory retrofitting of ceiling insulation in homes and rental units upon sale or reconnection of utility service. It estimates that uninsulated single-family dwellings still represent up to 42% of the stock. The uninsulated stock is primarily in southern California, however, where interstate gas prices have been highly regulated and maintained at an artificially low level. In northern California, where residents had been using higher-priced Canadian natural gas, fully 80% of the stock is estimated to be adequately insulated (Mr. Gordon Gill, California Energy Commission, Interview, October 1979). The commission has set a program goal of retrofitting 90% of the 4 million uninsulated and underinsulated dwellings over a five year period. As a further incentive to homeowners, the commission has recommended that utilities absorb one half to one third of the cost of retrofit insulation and expense it to operations.

Turning responsibility for energy conservation programs over to the utilities in California has proved to be a workable approach. Utility service representatives have an average of two contacts a year with each customer, in addition to monthly billings, and are in an excellent position to understand the needs of their customers. It was feared that conservation was counter to utility self-interest, but this fear has proved to be entirely unfounded in California.

6.8 CONCLUSION

In the early 1970s, California faced the intolerable proposition that its need for new electrical capacity would double every 10 years. Energy conservation regulations—including efficiency standards for refrigerators, freezers, room air conditioners, and water and space heaters —plus the state's solar and insulation programs will result in energy savings equivalent to 50 million barrels of oil by 1985 and five or six large power plants by 2000. These estimates do not include the effects of energy price increases, load management, marginal cost pricing of electricity, and the ultimate deregulation of oil and natural gas contemplated by the National Energy Act.*

*The National Energy Act is composed of five individual pieces of legislation: The Public Utility Regulatory Policies Act (P.L. 95-617), The Energy Tax Act (P.L. 95-618), The National Energy Conservation Policy Act (P.L. 95-619), The Powerplant and Industrial Fuel Use Act (P.L. 95-620), and The Natural Gas Policy Act (P.L. 95-621).

The conclusion to the CEC-sponsored study (CEC, 1979d, p. 37) of the cost competitiveness of solar and other forms of energy asserted that "generally it would be preferable to eliminate the present subsidies to gas and electricity as well as solar energy and to impose regulation or taxation upon conventional energy use which reflects health and environmental cost." In the case of California's 55% tax credit, as well as other historical subsidies to energy-producing industries, the income distribution effects are such that they benefit upper-income groups almost exclusively. If half the cost of new gas and electricity supply in California will be paid by tax and pricing subsidies to the utilities, as the study estimates, then expenditures on countervailing incentives and subsidies appear misspent. The use of economic incentives should be to bring the economic motivation and perceptions of individuals in line with the true social costs. As long as these costs are camouflaged, the incentive will be perceived only as a temporary gratuity.

7

ECONOMIC INCENTIVES FOR ENERGY CONSERVATION IN OREGON— A CASE STUDY

7.1 INTRODUCTION

In the face of rising oil prices and capacity limitations in its hydroelectricity system, Oregon has been progressive in formulating energy conservation policies for the residential sector. By 1974 a clause was added to the state building code that made mandatory the weatherization of new homes in Oregon. In 1977 the legislature instituted a program of voluntary economic incentives encouraging weatherization and alternate energy devices.

This chapter makes a preliminary assessment of these incentives in terms of participation rates, estimated energy savings, and the impact of implementation procedures on program effectiveness. The success of these voluntary programs appears to hinge on their degree of acceptability to the general public. As can be expected, programs requiring a minimum of financial outlay and inconvenience on the applicant's part are the most acceptable.

Before evaluating the programs, some background information is presented as well as an overview of the energy supply and demand conditions in Oregon. Subsequent to the analysis of the programs, Pacific Power & Light's experience with its interest-free, deferred-payment loan for weatherization program is analyzed. Finally, some comments are made on the city of Portland's recent ordinance requiring insulation of every home beginning in 1984.

7.2   ENERGY OVERVIEW FOR OREGON

7.2.1   Energy Supply

A major proportion of Oregon's energy requirements is imported. The state has no crude oil, natural gas, coal, or uranium reserves. The only major energy form that it produces is hydroelectric power and most of the sites with large potential have already been developed. Oregon has tapped a less conventional energy source, geothermal energy, but its use is limited to heating in the Klamath Falls area.

The following table shows the approximate proportion of Oregon's energy usage accounted for by each of the three major energy forms in 1980 (ODOE, 1982a):

Petroleum	58%
Electricity	22%
Natural gas	15%
Coal and other	5%

Oil

Oregon relies on petroleum for most of its energy needs. Crude oil is imported from Alaska and foreign sources, so the state can do little to influence petroleum availability or prices. Oregon's only refinery is principally for the production of asphalt; most petroleum products come from refineries in the Puget Sound area of Washington state.

Electricity

Most of Oregon's electricity is generated by hydroelectric projects. Because of the state's dependence on hydro resources, it is particularly susceptible to power shortages caused by drought conditions. These shortages can be alleviated by thermally generated power, but this source is much more costly than hydro. Oregon's hydro system is almost

TABLE 7.1.   Electrical Generation Facilities in Oregon, 1977

Source	Capacity (MW)	Average Megawatts of Output[b]	Percent of Total Output[e]
Hydro			
Federal[a]	3255	1505	51
Portland General Electric	518	226	8
Idaho Power	471	217	7
Pacific Power & Light	327	148	5
Miscellaneous[c]	148	69	2
Total	4719	2165	73
Nuclear			
Portland General Electric (Trojan)	1130	741[d]	25
Oil and gas			
Portland General Electric	932	11	4
Pacific Power & Light	36	—	
Biomass			
Eugene Water and Electric Board	85	30	1
Total	6902	2947	

Source: ODOE, 1979d, p. 9.

[a]One half of the capacity and output of facilities located in Oregon and adjoining states are included.

[b]Some 1977 figures are estimated.

[c]Small facilities are not well accounted for.

[d]1977 output only.

[e]Percentages do not sum exactly to 100%.

fully developed. While its peaking capability can be increased by install-ing additional generators at existing dams, the system's total energy out-put cannot be increased significantly.

Most of Oregon's electricity comes from the regional Northwest Power Pool, which coordinates the production and use of electricity generated by 109 publicly owned utilities and eight investor-owned utilities. The Bonneville Power Administration (BPA) and the public utilities distribute about one third of the electricity used in Oregon, although most is produced outside of the state. Private utilities distribute about two thirds of Oregon's electricity, producing about one half of that amount in state. Table 7.1 shows the breakdown of electricity production by source and type of fuel.

Portland General Electric operates Oregon's only nuclear plant (Tro-jan). Designed to operate at 70% of capacity, this plant operated at 66%

after its first full year of operation in 1977, and produced 741 MW of electricity, or 25% of Oregon's electricity demand. Two additional nuclear units are planned for installation near Arlington. They will come on stream in the late 1980s and each has a planned capacity of 1260 MW.

Oregon has about 1000 MW of oil- and gas-fired electrical generating capacity. This method of generation is costly because of the high price of fuel, but it is an effective and economic way of meeting peak demands and providing an alternate source of energy during electricity shortages caused by drought.

Coal is used to generate electricity in only one plant in Oregon, Portland General Electric's Boardman plant. It has a capacity of 530 MW. The fuel for this plant comes from Wyoming via unit trains.

Even though wind has been a source of energy for electrical generation in Oregon in the past, only a few wind-powered generators remain. Oregon is well-located for this type of power because of its proximity to the ocean and mountains where winds are significant. The major obstacle, however, is wind variability. This energy may not be available when it is needed most. Wind-powered generators have potential, however, because electricity storage problems are minor if the generators are part of a large mixed network of hydro and wind-powered sources. In addition, it is fortunate that wind energy is available in high-demand months, during the late fall, winter, and early spring.

Wood waste and sawdust are used to produce about 85 MW of electricity in the state. More extensive use of biomass is limited by its widely varying prices and lack of availability near generating facilities. If the prices of nonrenewable resources rise relative to biomass, then this renewable fuel will become more attractive.

A major problem facing Oregon has been how to allocate the existing low-cost hydroelectricity. Historically, publicly owned utilities and certain industrial customers have had access to low-cost federal power. Privately owned utilities have had to build more expensive thermal generation facilities and charge higher rates. Because only 20% of Oregon residents are served by public utilities, this rate disparity has become a major issue. Under provisions of the Northwest Power Planning Act (1980), the private utilities are permitted to sell their average cost power to the Bonneville Power Administration (BPA) and in return receive an equal amount of lower-cost hydroelectric power. In 1985 the amount of this power swap will be limited to the quantity used by the residential customers of the private utilities. In years prior to that date, proportionally smaller amounts will be swapped, based on an initial rate of 50% of residential demand in 1980. This will have the effect of giving the benefits of federal power to Oregon residential customers who are principally served by private utilities. On the other hand, public utilities and industrial customers served directly by BPA are facing rapidly increasing rates. The problems of the direct service industrial customers (DSIs) are particularly acute because most have had long-term power contracts expire recently, and most are in industries, such as aluminum production, that are very sensitive to electrical costs. BPA must allocate power in such a way that these industries, which are major employers, remain viable for the most part, while allowing BPA to recover the higher costs caused by the power swap. The consumption by direct service industrial customers accounts for 25% of the power produced in the Columbia River hydroelectric system administered by BPA.

The *Northwest Power Planning Act* mandates that BPA must acquire conservation and renewable resources to reduce load and the need for new generating facilities. In fact, because of currently falling electrical demand in Oregon, a projected low growth for electricity in the future, and generating plants under construction, there is likely to be no need for further capacity for 10–15 years. BPA plans a strategy of gradually adding conservation measures in order to moderate load growth as the need arises and, in the near term, concentrating on demonstration projects, research, and coordination of conservation programs offered by the utilities.

Natural Gas

Approximately two thirds of the natural gas used in Oregon originates in Canada, while the remainder comes from domestic production in the lower 48 states. Supplemental supplies are provided by LNG (liquefied natural gas) and coal gas. Canadian gas is based on long-term contracts.

The Northwest Pipeline Corporation uses two pipelines to import Oregon's Canadian gas supply from British Columbia and Alberta for distribution by three privately owned gas utilities. Supplies obtained during the low-demand periods are compressed into liquid form and stored at facilities in Portland and Newport until they are needed to meet peak demands during the heating season. Gas sales for 1981 by the three privately owned utilities are presented in Table 7.2.

TABLE 7.2. Natural Gas Sales by Privately Owned Gas Utilities in Oregon, 1981

Utility	Therms of Gas Sold	Percentage of Total Gas Sales
California-Pacific Utilities Company	67,852,367	8
Cascade Natural Gas Company	50,448,933	6
Northwest Natural Gas Company	751,162,424	86
Total	869,463,724	

Source: Oregon, PUC, 1982, pp. 55–57.

Prospects for Oregon's future gas supplies are reasonably good. Current arrangements are adequate to meet natural gas demand, which has declined somewhat since 1973. The state produces 3–4% of its own requirements from fields near Mist.

Geothermal Energy

Geothermal energy is indigenous to Oregon and has been used since the turn of the century for nonelectric applications. Although its contribution to the state's total energy supply is small, it is growing steadily and shows a great deal of potential. The U.S. Geological Survey estimates that "there is enough energy in Oregon's identified geothermal areas to produce 1640 MW of electricity and the equivalent of 9240 MW of direct process heat for a thirty-year period" (ODOE, 1979d, p. 39).

Solar Energy

In 1978 the Oregon Department of Energy conducted a study to determine the potential for solar power in the state. It found that solar systems can economically provide between 25 and 75% of space and water heating needs for many homes. Since a major determinant of the attractiveness of a solar system is economics, Oregon has implemented legislation that makes these installations more economically viable.

7.2.2 Energy Demand

Energy demand by sector, as estimated by the Oregon Department of Energy, is shown in Table 7.3. Important shifts have occurred in the

TABLE 7.3. 1980 Oregon Energy Demand by Sector

Sector	Use (trillion Btu's)	Percent of Demand
Residential	103.6	19.3%
Commercial	65.6	12.2
Manufacturing	123.9	23.1
Transportation	230.1	42.9
Agricultural and others	12.9	2.5
Total	536.1	100.0%

Source: ODOE, 1983d.

TABLE 7.4. Energy Demand by Fuel in Oregon (trillion Btu's)

Year	Electricity	(%)	Natural Gas	(%)	Petroleum	(%)	Other
1971	96.2	21	89.9	19	275.6	60	NA
1974	110.9	23	95.4	20	281.6	57	NA
1975	114.0	24	86.8	19	267.2	57	NA
1976	120.8	25	86.5	18	279.2	57	NA
1977	122.5	24	90.7	18	298.9	58	NA
1978	126.6	23	88.6	16	324.4	61	NA
1979	134.6	25	92.4	17	321.5	58	NA

Source: ODOE, 1983d.
Note: NA, not available.

residential and manufacturing sectors during the 1970s, resulting in an overall declining share for petroleum in total energy consumption (Table 7.4). At the same time, electricity has increased in importance during the decade.

Total residential demand for energy in Oregon increased 3.8% from 1971 to 1979 (Table 7.5). During this period, demand peaked in 1972 and then declined until 1975, mostly because of the impact of oil shortages and price increases in the 1973–1974 period. In 1975, residential energy demand began increasing again.

The composition of this demand has changed significantly over the decade. In 1971, oil and electricity accounted for 30% and 36.2% of residential usage, respectively. By 1979, however, oil accounted for only 17%, whereas the share of electricity had grown to about 50%. The share of natural gas has remained relatively stable, although it showed a decreasing trend from 1975 to 1979.

TABLE 7.5. Residential Energy Consumption in Oregon (trillion Btu's)

Year	Electricity (%)	Natural Gas (%)	Oil[a] (%)	Other	Total
1971	36.2	22.4	30.0	NA	88.6
1974	40.0	22.8	25.9	NA	88.7
1975	42.7	23.4	24.2	NA	90.3
1976	43.7	21.8	22.7	NA	88.2
1977	44.7	21.0	21.0	NA	86.7
1978	46.4	20.6	19.3	NA	86.3
1979	49.6	21.1	16.9	NA	87.6

Source: ODOE, 1983d.
Note: NA, not available.
[a]ODOE estimate.

**TABLE 7.6.   Transportation Energy Consumption
in Oregon (trillion Btu's)**

Year	Electricity	Petroleum	Total
1971	0.5	182.4	182.9
1974	0.6	197.8	198.4
1975	0.6	201.9	202.5
1976	0.6	214.9	215.5
1977	0.7	230.0	230.7
1978	0.7	253.9	254.6
1979	0.7	260.7	261.4

Source: ODOE, 1983d.

Energy demand in the transportation sector is illustrated in Table 7.6. Oregon's per capita petroleum usage is slightly less than the U.S. average, but its trend parallels that of the nation.

From 1973 to 1976, Oregon's total per capita energy consumption was slightly lower than the national average (ODOE, 1979d, p. 44). It would be wrong, however, to assume that energy was used more efficiently in Oregon than in the United States as a whole. There are many factors that could explain this discrepancy, including economic structure, weather, and population density. Furthermore, if Oregon's use of such fuels as firewood and biomass could be accounted for more accurately, per capita energy usage might be much closer to the national level. Based on energy demand trends for Oregon and the United States over the 1973–1976 period, the state's Department of Energy has concluded that "before Oregon implemented major energy conservation legislation [1977], the state's energy conservation performance was neither significantly better nor worse than the national average" (ODOE, 1979d, p. 45).

7.3   STATE LEGISLATION

7.3.1   Background

In 1977 the state of Oregon passed enabling legislation for several specific energy conservation programs. A major stimulus for this action was the passage of the Energy Policy and Conservation Act (EPCA) by the federal government in 1975. The purpose of EPCA was to provide legislation that would restrain energy demand and increase domestic supplies and availability, as well as deal with emergency situations such as an oil shortage.

A major component of EPCA was a provision for the development and implementation of state energy conservation plans. The federal government offered financial and technical assistance to the states for developing specific programs that would promote energy conservation and reduce the rate of growth of energy demand. The specific goal was "a reduction, as a result of the implementation [of] the State energy conservation plan . . . of 5 percent or more in the total amount of energy consumed in such State in the year 1980 from the projected energy consumption for such State in the year 1980" (section 362, EPCA, 1975).

Amendments were made to EPCA by the Energy Conservation and Production Act (ECPA) passed in 1976 and the National Energy Act (NEA) passed in 1978. ECPA offered financial assistance to the states for the implementation of a weatherization assistance program for low-income persons, particularly those that are elderly or handicapped. It was recognized that low-income members of society are those least able to afford weatherization for their homes. Such a program would not only decrease the energy bills for these people but also conserve energy. This amendment led to the implementation of Oregon's Senate Bill 4 in 1977, or the Elderly Low-Income Weatherization Program. ECPA also required that a supplemental state energy conservation plan be added to the plan already described in EPCA.

The National Energy Act extended federal financial assistance through fiscal year 1979 for state programs implemented under EPCA and ECPA. Financial aid for the weatherization program was extended through fiscal year 1980. In order to be eligible, states were required to undertake specific conservation actions, as they had been under EPCA. These requirements are discussed in more detail later in this chapter.

7.3.2 Oregon Energy Conservation Plan

The state of Oregon responded to EPCA and ECPA by developing the Oregon Energy Conservation Plan. A number of programs were implemented to achieve the goal of reducing the state's projected 1980 energy consumption by at least 5%. These programs were deemed to be consistent with energy demand and supply conditions in Oregon and constraints of technological feasibility, financial resources, and economic objectives. It was intended that the programs would "reduce overall consumption of energy" and "develop those sources of energy which are available in [Oregon]" (ODOE, 1979a, p. 13). The criteria used in program selection were "that there be a measurable energy savings by 1980 and that there be a continuing energy savings beyond that time."

EPCA and ECPA included a number of mandatory requirements in order for a state to be eligible for federal financial assistance. EPCA required that state plans include the following components:

Mandatory lighting efficiency standards for public buildings.

Programs to promote the availability and use of carpools, vanpools, and public transportation.

Mandatory standards and policies relating to energy efficiency in the procurement practices of the states and their political subdivisions.

Mandatory thermal efficiency standards and visualization requirements for new and renovated buildings.

A traffic law or regulation that, to the maximum extent practicable consistent with safety, permits the operator of a motor vehicle to turn the vehicle right at a red traffic light after stopping.

Legislation was passed in Oregon in 1977 to satisfy the lighting standards requirement. The procurement practices of the state of Oregon had included energy efficiency provisions since 1975. Similarly, the Oregon Uniform Building Code (Chapter 53, 1974) had already established thermal efficiency standards for all new residential buildings, although these standards have been further upgraded.

The two requirements relating to transportation have also been met in Oregon, and the state Department of Energy reports the operation of the following programs (ODOE, 1979a, p. 15):

Carpool: administered by Tri-Met in the City of Portland and designed to match riders to promote carpooling.

Commuter Club: a bus pool in operation since 1974 that carries approximately 75 people per day between Portland and Salem.

Over 78 Park and Ride lots in the Portland metropolitan area for transit riders.

Specially designated carpool, vanpool, and bus lanes on Portland's Banfield Freeway.

Bikeways funded through 1% of the state's revenues on gasoline taxes.

A shuttlebus serving state employees traveling on official business between the three major metropolitan areas.

ECPA required Oregon to undertake four additional activities (ODOE, 1979a, p. 15).

A public awareness program publicizing the availability of energy audits.

Public education with respect to planning, financing, installing, and monitoring the effectiveness of energy conservation measures.

Procedures to provide for effective coordination of existing local, state, and federal agency conservation programs.

Procedures for encouraging and carrying out energy audits.

All of these requirements have been met. Details about the energy audits are provided later in this chapter.

Projected energy savings and costs of implementation of the Oregon Energy Conservation Plan are summarized in Table 7.7. The plan incorporates 14 program categories that together were projected to save about 68.8×10^{12} Btu of energy by 1980, or 14% of 1977 total energy demand in Oregon, and cost approximately $963,000 to implement in that same year. The largest portion of the 1980 budget, or 27.5%, was directed toward renewable resources. In contrast, the Residential Energy Conservation Program was allocated only 3.9% of the 1980 budget, yet was predicted to generate the most energy savings, that is, about 16.5×10^{12} Btu, or 3.3% of Oregon's total demand in 1977 (ODOE, 1979d, p. 42). The projected energy savings of the Residential Energy Conservation Program represents approximately 20% of total demand in this sector in 1977 (ODOE, 1979d, p. 47). This case study will focus on the effectiveness of the program measures undertaken in the Residential Energy Conservation category.

7.3.3 Analysis of Residential Energy Conservation Programs

The specific residential energy conservation programs are presented in Table 7.8 together with their respective bill numbers and projected energy savings. Enabling legislation was passed by the state in 1977. It is possible to make a preliminary assessment of participation rates, estimated energy savings, and the impact of implementation procedures on program effectiveness.

All the programs are voluntary economic incentives aimed at either encouraging weatherization or the installation of alternate energy devices.

Personal Income Tax Credit Programs

WEATHERIZATION TAX CREDIT (HB2701)—1977

*Description.** House Bill 2701 allowed a personal income tax credit for individual taxpayers who weatherized or who otherwise improved the energy efficiency of their principal residence or the principal residence of their renters, excluding mobile homes. The credit could not

---

*ODOE, 1977b. This source is used for many of the ensuing program descriptions.

TABLE 7.7. Projected Savings of Oregon's Residential Energy Conservation Programs

Program Measures	Gross Estimated Energy Savings (in Btu's)^a		Cost of Implementation		Percent of 1980 Energy Conservation Budget
	By 1980	Annual 1981	Estimated Total to 1980	Annual 1981	
1. *Residential energy conservation*	16.522×10^{12}	5.186×10^{12}	$120,297		3.94
Inverted utility rates				—	
Utility loans				—	
State weatherization tax credit				$ 605,097	
Senior refund				expired	
Low-income grants				n/a	
State veteran's loans				n/a	
State oil loans				$ 38,340	
Pilot light ban				—	
2. *Building codes*	10.566×10^{12}	0.833×10^{12}	$275,557	n/a	8.03
Residential					
Commercial					
3. *Public awareness*	11.4542×10^{12}	n/a	$282,265	n/a	9.45
4. *Local energy program*	1.6244×10^{12}	n/a	$572,809	n/a	19.33

5. Education	3.0405×10^{12}	n/a	$151,992	n/a	6.01
6. Transportation	7.2793×10^{12}	0.486×10^{12}	$136,809	n/a	1.09
7. Community energy planning	2.7954×10^{12}	n/a	$139,181	n/a	5.88
8. Commercial/industrial conservation	10.8023×10^{12}	$1.5 \ \times 10^{12}$	$230,645		10.04
Small-scale energy loans				$ 450,000	
Business/industry tax credit				$ 400,000	
9. Energy audits	0.5184×10^{12}	n/a	$100,478	n/a	n/a
10. Agriculture	$0.003 \ \times 10^{12}$	n/a	$ 24,424	n/a	n/a
11. Purchasing	0.1808×10^{12}	n/a	$ 34,132	n/a	n/a
12. Energy management in state facilities	2.8732×10^{12}	1.439×10^{12}	$261,826	n/a	8.69
SEMP/St. Bldg. Fund					
Institutional buildings fund					
13. Renewable resources	1.5514×10^{12}	0.037×10^{12}	$789,388	$2,600,000	27.53
State renewables tax credit					
14. Assessments			$ 49,846	n/a	n/a

Sources: ODOE, 1979a, pp. 11–12; 1979d, p. 42; 1983d; and various internal memos.

[a]Total net energy demand in Oregon in 1977 was 496.7×10^{12} Btu's.

TABLE 7.8. Residential Energy Conservation Programs in the Oregon Energy Conservation Plan

Program	State Bill Number	Estimated Energy Savings by 1980 (in Btu's)
1. Personal income tax credit programs		
Weatherization tax credit	HB2701	6.0643×10^{12}
Alternate energy device tax credit[a]	SB339	—
2. Veterans' loan programs		
Loans for purchase of weatherized homes or retrofitting of existing homes	HB2156	2.5988×10^{12}
Alternate energy device loans	SB477	0.0
	HB2253	?
3. Energy supplier conservation services programs		
Publicly-owned utilities' services	HB3265 ⎫	
Investor-owned utilities' services	HB2157 ⎬	6.6284×10^{12}
Energy suppliers' services	SB371 ⎭	
4. Elderly low-income weatherization refunds	SB4	1.2307×10^{12b}
Total		16.5222×10^{12}

Source: ODOE, 1979b, Appendix A.
[a]SB339 falls within the Renewable Resources Program category.
[b]This saving includes 0.1170×10^{12} Btu under SB4 and 1.1137×10^{12} Btu under a federal program.

exceed the lesser of $125 or 25% of the actual cost of purchasing and installing such items as caulking, weatherstripping, insulation, vapor barrier materials, timed thermostats, dehumidifiers, and storm windows and doors.

A taxpayer could claim only one credit per year and must own the dwelling to be eligible. The full amount of the credit could be claimed on a dwelling only once during the lifetime of the law (1977–1984) unless there is a change in dwelling ownership. The new owner would have been eligible for a similar tax credit for additional weatherization that he or she installed. If the amount of the credit exceeded the taxpayer's liability, all or part of the balance could be claimed within five successive years until it was fully used.

Weatherization items must have been installed in the dwelling after October 4, 1977, but before January 1, 1981. As proof of installation, the taxpayer had to submit a statement with his or her tax forms certifying that the materials and the installation met the Oregon Uniform Building Code. This statement had to be signed by an insulation or general contractor who was registered with the state Builder's Board or signed by a building official or inspector who was certified under Oregon law (this requirement was dropped early in the program).

If the taxpayer was receiving any amount of weatherization assistance, aid, grant, refund or subsidized (low-interest) loan, he or she was not eligible for a tax credit. However, a veteran who received a loan for weatherization under HB2156 could also receive a tax credit under HB2701. Weatherization costs could not be added to the basis of property value for depreciation purposes or for the computation of gain or loss on sale or other disposition of property. In 1981 the legislature phased out this program retroactively to January 1, 1981, except for oil-heated homes, which were "sunsetted" on September 1, 1981.

Response. Compared to projections made by the Oregon Department of Energy (ODOE), the program was quite successful. In 1978 ODOE estimated that 30,000 of the state's residential taxpayers would take advantage of the tax credit by 1980, resulting in a total energy saving of approximately 3.7×10^{12} Btu (ODOE, 1978a, p. A23). This response would represent about 3.3% of Oregon's 1977 stock of single-family residences and apartments. In fact, 32,757 tax credits were claimed for 1978 and 70,000 were claimed for the 1979 tax year (Oregon Department of Revenue). In 1979 the ODOE revised its anticipated energy savings by 1980 to about 6.06×10^{12} Btu, reflecting the greater than expected response to the program, together with changes in some of the underlying assumptions of the projections (ODOE, 1979b, Appendix A, p. 33).*

Evaluation. The Weatherization Tax Credit, like other energy conservation incentive programs in Oregon, was implemented to encourage more efficient energy use, to decrease homeowner costs, and to make residential energy available for other uses. Depending on the age of a residence and the level of weatherization, a consumer could use this program to help decrease home energy consumption by as much as 40%. As an illustration of the benefits of such programs, if all the homes in Oregon were properly weatherized, the equivalent of about 75% of the Trojan nuclear plant's output would be saved. Obviously, any amount of weatherization installed, which is even reasonably cost effective, is a significant increment to achieving the program's goals.

Before this program became available, there was already a general interest in weatherization activities. The program provided the necessary push for those people who had been considering weatherization but

*ODOE originally assumed that the energy savings resulting from the weatherization program would be 30% of the average amount of energy consumed for space heating for each participating household. When the 1979 revisions were made, this assumption was changed to 10%.

had not taken any action. The tax credit was a strong economic incentive. For example, a $125 tax credit represents 25% of a $600 expenditure. Not only was the capital outlay for the homeowner reduced but the state assisted him in achieving energy cost savings over time. Furthermore, the value of weatherization materials is not assessed for property tax purposes, thereby avoiding a "built-in" disincentive that might otherwise reduce the effectiveness of this program. Even during an inflationary period, the available tax credit was a strong incentive.

The total value of credits claimed in 1978 was nearly $3 million, or $89 per claimant (Oregon Department of Revenue, 1980). According to an ODOE follow-up survey using a random sample of 450 homes, the mean cost of weatherization was $729. The most popular weatherization activities were of the "do it yourself" variety and included insulation, storm doors and windows, and glass fireplace screens.

Because the tax credit was available to all homeowners, it had the potential to reach many more people than other state programs. To ensure that the public was aware of the tax credit opportunity, the Oregon Departments of Energy and Revenue implemented a "very heavy" public information campaign. Radio, television, newspapers, pamphlets, and speeches were utilized. By August 1979, $20,000, or 44% of the combined 1977 and 1978 budgets for the residential energy conservation program category (Table 7.6) had been spent on promotion using the media. A further $25,000 was spent to produce 150,000 copies of each of two ODOE brochures. They were distributed at many types of outlets, including liquor stores and banks, and about 6000 had been mailed by August 1979 in direct response to public inquiries. The information program emphasized that energy costs were rising and that money could be saved by weatherizing.

The Department of Revenue, which implemented the tax credit, also had an information program, thus lowering combined program costs because of institutional mechanisms already in existence. One employee had responsibility for the dissemination of information and she spent approximately 10% of her time publicizing the energy conservation tax credits by distributing circulars, preparing news releases, contacting the press and general public. In addition, a "tax help unit" fielded toll-free telephone inquiries concerning eligibility criteria and other program details. Direct costs were limited to postage and the printing of 20,000 information circulars. Although the "save money" theme was used occasionally in news releases, the department kept a low profile and was more of an information provider than an advocate of the tax credit.

A major problem with House Bill 2701 was determining what qualified as "weatherization." Despite comprehensive lists of eligible and ineligible items in the information circular and brochure, there were many inquiries about the eligibility of unlisted items. Early in the program, certain weatherization materials had to be installed to meet the requirements of the Oregon state building codes in effect at that time. Compounding the qualification problem was the fact that the list of qualifying items and the building codes kept changing, so that a person weatherizing in the hope of obtaining a tax credit might have incorrect information. The requirement for a signed certificate indicating compliance with the building codes was eventually dropped because the city inspectors and contractors who were registered with the Builder's Board were too busy.

The weatherization industry is easy to enter and, when House Bill 2701 increased the market for weatherization materials, some "not quite reliable and responsible" insulators appeared on the scene. No outright fraud leading to prosecution had occurred as of August 1979, but the Federal Trade Commission brought action against one manufacturer whose insulation, not treated with proper fire retardant, was involved in three fires.

The weatherization tax credit was made available about the same time that investor-owned utilities began to offer interest-free loans for weatherization (see section on House Bill 2157). Had the state legislature been aware of these loans when the legislation was drafted and considered, the bill might have been modified. Ostensibly, the loans were competing against the tax credit; however, the tax credit still served a useful purpose. In 1979 the investor-owned utilities had a huge backlog of requests for their weatherization services, so the tax credit provided Oregonians with a weatherization alternative that could be implemented faster.

ALTERNATE ENERGY DEVICE TAX CREDIT (SB339)—1977

Description. Senate Bill 339 amended earlier programs that offered a tax credit for a solar energy device and that exempted solar systems from property taxation. It provides a tax credit to any Oregon homeowner who installs solar, wind, or geothermal energy equipment in his principal or secondary residence. Twenty-five percent of the investment cost (or a maximum of $1000) may be claimed, provided the device meets minimum performance criteria and has been certified by the Oregon Department of Energy.

The performance criteria originally included the requirement that the device generate at least 10% of the dwelling's total energy need. This criterion was later amended to allow also for systems that met 50% of the dwelling's hot water requirements. Taxpayers are eligible for only one credit per year and must claim it during the year that the equipment has been certified. If the amount of the credit exceeds the taxpayer's liability, the credit may be claimed for five successive years until it is fully used. The program took effect January 1, 1978.

The value of a solar energy heating or cooling system is exempt from ad valorem property taxation, and the exemption applies to any installation made on or after January 1, 1976, but before January 1, 1998. Swimming pool heaters providing 10% of the dwelling's total energy requirement may also, on certification, qualify for the tax credit.

Response.   Participation in the program has exceeded the projections made by the ODOE before its implementation. At that time, it was estimated that 300 individual taxpayers would claim the credit between 1977 and 1980. Under the ODOE's assumption that the average energy savings resulting from the installation of an alternate energy system would be 50% per home, the department projected a total saving of 0.0619×10^{12} Btu by 1980 (ODOE, 1978a, p. A137). As of December 31, 1981, there were 13,800 applications on a preliminary basis, of which 99% were for solar equipment. Eighty percent of these devices are used in hot water applications rather than space heating.

According to the ODOE, the overwhelming response to this program has been the result of the changed criteria for performance and the change in the federal tax credit for solar devices from 15 to 40% in April 1980.

Evaluation.   The bulk of the certifications under the program has been for solar energy equipment. The relatively large number of these installations is significant, considering that a few decades ago it was thought that the Oregon climate was unsuitable for this activity. The fact that only 12 such systems existed at the beginning of the program illustrates the previous lack of interest in solar energy. It was not until the ODOE showed that this type of power was more cost effective than previously recognized and until the 25% tax credit became available that interest was triggered, resulting in the rapid development of Oregon's young solar industry.

A key element of the program is the application for certification, which is a preliminary step for receiving the state tax credit. The certification questionnaire is designed to be filled out by the applicant, with no ODOE on-site inspection required. A sun chart and necessary information regarding location and structural parameters of the house as well as the specifications of the proposed system are required. An alternative method to apply for the tax credit is to use an ODOE certified dealer. About 60 dealers were certified by 1982, and 20% of the tax credit applications are filed by that process. A summary of the Renewables tax credit program is shown in Table 7.9.

The typical applicant has a number of general characteristics. He or she is about 24–40 years old and has a "decent" income, because most systems are expensive. The average tax credit in 1978 was $670, implying a total installed system value of at least $2680. The applicant is mechanically and artistically inclined because many people design and

TABLE 7.9.   Oregon Residential Renewable Resource Tax Credit Program Through December 31, 1982

Year	Applications Received	Cost Incurred to General Fund (millions $)	Energy Savings (MWh/year)	Cumulative Energy Savings of Installed Projects (average MW)
1977	27	0	—	—
1978	163	—	—	—
1979	414	0.188	—	—
1980	2510	1.27	—	—
1981	5968	2.6	13,410	1.5
1982	5000	3.1 (est.)	10,180	2.7
1983	7500 (est.)	3.5 (est.)	11,280	4.0
1984	8400 (est.)	4.04 (est.)	12,650	5.4

Total cost estimates	$14.7 million
Effects of cumulative energy saving	5.4 av. MW
Cost/savings	$2.72/av. watt

Source: ODOE, 1983a.

install their own equipment, and commercially produced systems are not very easy to install. The applicant usually has some facility with aspects of taxation because he or she must recognize the availability and desirability of a tax credit and how to obtain one. The applicant supports environmental protection or has a very high energy bill. From this profile it appears that the major hurdles to the acceptability of such a program are the high start-up costs and the mechanical complexities of alternate energy systems. Based on the large number of inquiries that the ODOE receives, there is definitely an awareness about the program among Oregonians. Promotional material is incorporated in information disseminated about other state energy activities.

There was a major problem that occurred when the program was first implemented. The bill initially provided for the tax credit to be retroactive for some time period before October 4, 1977. Publicity was generated in the spring of that year, which resulted in a number of people installing systems without prior certification. Unfortunately, the retroactive clause was removed from the bill later, and those who had installed their systems before October 4, 1977, could not receive the tax credit. Even if the legislation had been retroactive, however, the credit could not have been given for systems without prior certification. Careful planning and a good information program could have avoided these problems.

In summary, the alternate energy tax credit is effective. The participation rate and the potential for energy savings are higher than projected.

Awareness of the program is high, although participation might be inhibited by the actual or perceived technical knowledge required.

Veterans' Loan Programs

LOANS FOR PURCHASE OF WEATHERIZED HOMES OR RETROFITTING OF EXISTING HOMES (HB2156)—1977

Description. House Bill 2156 requires that in order for a veteran to be eligible for a loan from the Oregon Department of Veterans' Affairs (ODVA) for the purchase of a home built prior to July 1, 1974 (when state insulation standards went into effect for newer homes), the home must meet minimum weatherization standards. A veteran may add the cost of weatherization to the principal of the loan and the weatherization standards must be met within 120 days of the issuance of the loan.

The Oregon Department of Commerce, in consultation with the Energy Conservation Board, is responsible for the weatherization standards.* The home must have weatherstripping around the doors and windows. If it is cost effective there must be R30 insulation in the ceiling with proper allowance for ventilation, R19 insulation under the floors that are over unheated spaces, R7 with appropriate protective covering around supply and return heating air ducts in unheated spaces, tightly fitting storm windows or insulating double glazing, R11 in the walls, and R11 with appropriate protective covering around hot water heaters located in unheated spaces. *Cost effective* means that the energy saved by the weatherization in dollars must be less than the increase in mortgage payment due to the weatherization at the ODVA interest rate prevailing at the time. *Unheated space* means any place exposed to ambient temperature and not provided with a heat supply capable of maintaining a minimum temperature of 15.5°C (60°F).

The legislation also provides for loans for home improvements to meet weatherization standards. House Bill 2156 became effective on October 1, 1977.

Response. Participation in the loan program is greater than was projected in April 1978. About 28,000 homes had been weatherized by August 1979, compared to the projection of 22,155 for the 1977–1979 period (ODOE, 1978a, p. A24). The original projection was based on the assumption that 80% of the total number of loans processed by the Oregon Department of Veterans' Affairs (ODVA) would be affected. Energy savings by 1980 were estimated to be 1.5068×10^{12} Btu, based on a 30% increase in thermal efficiency for each home weatherized. This estimate

*Subsequent to the formulation of the standards in 1977, the Energy Conservation Board was dismantled and its duties transferred to the Oregon Department of Energy.

was revised in 1979 to a saving of 2.5988×10^{12} Btu after an analysis of the greater than expected participation.

Evaluation.   The ODVA is a convenient vehicle for implementing the weatherization loan program. The program reaches many people, as the department processes 25% of all single-family residential mortgages in the state through 15 field offices and makes low-interest loans available to those residents who do not have access to the weatherization programs of the private utilities.

The ODVA's implementation costs have not been very high because the necessary mechanisms for a loan program were already in existence. The department had a staff of appraisers who could conduct the necessary home energy analysis. Some new administrative forms had to be developed and an information program launched. When House Bill 2156 was implemented, the department staff and general budget increased by about 10%, but a portion of this growth was not directly related to this loan program.

A weatherization loan has a number of advantages as an economic incentive. It is more attractive than a tax credit to the less "sophisticated" consumer. The effectiveness of tax credit programs can be inhibited by their tendency to intimidate those people who do not understand taxation principles. In contrast, most families are eventually faced with the task of negotiating a home purchase loan. By incorporating a weatherization loan into this process, it is possible to achieve energy conservation without having to "pull teeth." The ODVA loan program reaches a broad spectrum of consumers, including the low-income earners. At an interest rate of 5.9% (versus the market rate of about 11% in August 1979) and a 5% down payment requirement, ODVA loans have been very attractive in themselves. Not everyone is eligible, however, for these loans are restricted to war veterans and only two are allowed in a lifetime. About 45% of all eligible veterans have already applied for home loans through the department. By January 1982 the interest rate for ODVA loans had risen to 10.5% versus a market rate of 14.5%.

The information program is low key. At first, ODVA relied on word of mouth. At that time there was pressure put on the department by the general public, who perceived the weatherization requirement for a loan as another government attempt to "cram something down people's throats" and the creation of more bureaucracy. Since then, however, the appearance of other programs, especially those of the investor-owned utilities, and the growth in home energy bills have legitimized the ODVA initiative. Simultaneously, there have been more information demands on the ODVA. Fifty percent of the inquiries fielded by the department's architectural consulting arm, which is implementing the program, concern House Bill 2156. Public complaints now concern contractors who have failed to insulate up to standards. A number of community organizations have requested that the department sponsor speakers to discuss the loan program. The ODVA has provided further information in several of its pamphlets and the loan availability is also publicized in ODOE brochures.

Many of the ODVA speeches have been made to realty and contractor organizations that oppose the loan. These groups do not like the weatherization requirement because in many cases it delays their commission checks. Some contractors are also unaware of the technical requirements of proper weatherization. Consequently, the ODVA presentations must be persuasive and informational in content. The lesson from this experience has been that, before implementing similar programs, consultation is necessary with influential people in all groups who would be affected.

At the start of the program, the ODVA weatherization requirements were stiffer than those of the Oregon building code first introduced in 1974. Since then, however, the building standards have been upgraded, so that by 1979 they were stricter than those required by the ODVA. Oregon has found that the building code weatherization standards, which include R30 insulation in the ceilings, R19 in the floors, and R11 in the walls, can be made even more stringent without significant consumer resistance. The ODVA subsequently raised its standards to match those in the code. Furthermore, House Bill 2156 specifies that cost effectiveness calculations should be based on the home loan payback period, which can be as much as 30 years. Operationally, however, the ODVA originally required a payback period of 15 years on weatherization loans, making a lot of such activities cost ineffective. This provision was later revised to allow longer paybacks. The average value of a loan has been about $1290. Since R30 fiberglass batt had a retail price of about 30 cents per square foot in the U.S. Pacific northwest in 1982, a $1000 expenditure represents a significant injection of weatherization (price data from ODVA, January 1983). Some loans have been as high as $4000, whereas the small loans have typically been around $600.

A major problem with the program has been the consumer resistance to the installation of floor insulation. Since the loans are given directly to applicants, many do their own labor rather than contracting it out. Unfortunately, access for floor insulation is often difficult because of cramped crawl spaces. The job is awkward, uncomfortable, and dirty. Furthermore, the requirement is an extra financial burden. Alternative methods are to insulate the concrete wall around the perimeter of the crawl space or merely close the vents in the winter. The ODVA cannot recommend these alternatives, however, because they cause condensation, which leads to the rotting of the home's wood frame after about eight years. The department must protect the value of the home, if only for the purpose of loan security.

A major objective in designing a voluntary energy conservation incentive program is to make it easy for a consumer to participate. The ODVA loan is a significant step in achieving that objective. By combining the weatherization and home purchase loan, the costs of consumer efforts are minimized. The interest-free, deferred-payment loan programs of the investor-owned utilities have progressed even further in this regard because the utilities arrange for weatherization installation. These tactics for achieving consumer acceptability can be characterized as a "massage" approach. The consumer just lies back and has someone else do the work.

ALTERNATE ENERGY DEVICE LOANS (SB477)—1977

Description.   Senate Bill 477 applies to all veterans intending to install solar, wind, or geothermal energy devices in their homes. A loan of up to $3000 may be granted, provided the device will meet or exceed 10% of the total energy requirements of the home. Along with the Department of Veterans' Affairs, the Department of Energy establishes minimum performance criteria for such systems and uses these standards for certification. Veterans are also eligible to obtain a tax credit for alternate energy devices under Senate Bill 339. This program was amended in 1981 to increase the loan limit to $5000 under House Bill 2253.

Response.   Public response has been greater than projected by the ODVA in late 1978. At that time it was estimated that more than 24 alternate energy device loans would be approved in the 1977–1979 period (ODOE, 1979d, p. 34). Unfortunately, a separate statistical tracking of these loans was not made prior to July 1979. Instead, weatherization (HB2156) and alternate energy device loans (SB477) were lumped together. When separate accounting for Senate Bill 477 was started, however, it was found that 63 loans were made between July and December 1979 (ODVA correspondence, January 14, 1980). Hence, the response for those six months alone overwhelmed the total projection made for two years. The breakdown of the total number of devices installed was 55 solar, 6 geothermal, and 2 wind. Public response has been increasing such that in the first quarter of 1982, 206 loans for a total loan amount of $1,030,000 were made for renewables.

Evaluation.   Much of the discussion about House Bill 2156 also applies to Senate Bill 477, since the eligibility criteria and implementation procedures are practically the same. In August 1979, there were two loans available to veterans through the ODVA: $50,000 for those who have owned a home before and $58,000 for first-time home buyers. The maximum $5000 loan for an alternate energy device can be added to either of these loans. However, the recipient is permitted to spend as much of the combined amount as he wants on the device.

Table 7.10 presents a comparison of the responses to the tax credit and loan. There is a consistent proportional response across solar, geothermal, and wind equipment for the two types of incentives. Passive and active solar devices are by far the most popular, accounting for about 90% of all installations. The tax credits that are available for energy devices are much more attractive than the loan because the Federal Energy Tax Act of 1978 and Oregon Senate Bill 339 together can provide 55% of the cost. One advantage of the low-interest loan (5.9%), however, is that the down payment is eliminated and the loan can be repaid out of energy cost savings.

Senate Bill 477, together with Senate Bill 339, has created a market for alternate energy devices. Among the new entrepreneurs in the industry are a few of questionable competence and ethics. Some of the designs inspected by the ODVA have been very poor.

As a whole, the alternate energy device loan is not a strong component of Oregon's Energy Conservation Plan. The tax credits that are available seem to have provided a much more attractive incentive.

Energy Supplier Conservation Services Programs

PUBLICLY OWNED UTILITIES' SERVICES (HB3265)—1977

Description.   House Bill 3265 requires that publicly owned utilities and fuel oil dealers provide weatherization services to their space-heating customers. When this bill was passed in 1977, there were 31 publicly owned utilities and over 300 fuel oil dealers in Oregon. Provision of the following services are required on request:

Information about available weatherization programs.

Technical assistance concerning various methods of saving energy, including an inspection of the customer's home and a cost estimate for the installation of recommended weatherization (alternatively called a home energy analysis or audit).

A list of registered contractors providing various types of weatherization services within or in close proximity to the service area of the energy supplier.

Information about low-interest loan programs for weatherization available from lending institutions.

A "space-heating customer" is a dwelling owner or a tenant with at least three years remaining on his lease at the time that the weatherization services are performed.

TABLE 7.10. Comparison of Responses to Oregon's Alternate Energy
Device Tax Credit (SB339) and Alternate Energy Device Loan (SB477)

	SB339 October 4, 1977– September 19, 1979		SB477 July 1979– December 1979	
	Total Number	Percent of Total	Total Number	Percent of Total
Solar	438	92	55	87
Geothermal	23	5	6	10
Wind	16	3	2	3
Total	477	100	63	100

Source: Oregon Department of Energy.

The bill restricted the interest rate that a lending institution could charge for loans for weatherization to a maximum of 6.5% annually. The lending institution was given a tax credit, the value of which equaled the difference between 6.5% and a maximum 12% market annual interest rate that otherwise might have been charged for such loans. This bill is administered by the ODOE.

In 1981, legislation in House Bill 2247 revised the program, removing the 12% ceiling on the maximum annual market interest rate. The subsidized rate, however, remained at 6.5%. This change was necessitated by rising market interest rates that exceeded the cap of 12% and removed the incentive for commercial lenders to participate in the program. House Bill 2246 set up the ODOE as administrator of the oil heating customers' energy audit program. A $400,000 assessment against gross operating revenues of oil distributors funds the program, which the ODOE contracts out to Energy Counselors, Inc. The publicly owned utilities are required to perform energy audit services and provide cash incentives or 6.5% loans under 1981 legislation (Senate Bill 111). The cash incentives are 25% of the weatherization costs or $350, whichever is less.

Response. Public response to the original programs under House Bill 3265 was poor. The ODOE reported that 7226 homes, or about 2% of those eligible in 1978, were audited between October 4, 1977, and July 1, 1979 (Intergovernmental Relations Division, Oregon, correspondence, August 3, 1979). This compares to a projected 15% annual participation rate at the start of the program (ODOE, 1978a, p. A20).

In the period 1977–1981, 16,900 audits were completed for customers of publicly owned utilities and 1100 loans were made for weatherization. However, response was higher in the first nine months of 1982 with 8200 audits completed and 1458 cash grants financed. It is apparent that interest in these programs is much stronger because of the revised legislation. This is a combined effect of the revival of the 6.5% loan program and the new cash grant feature.

Evaluation.   An ODOE official wrote a memorandum for the state government in July 1979 evaluating the early results of Oregon's energy supplier conservation services programs. The analysis presented here is based on that memorandum.

Most of the publicly owned electric utilities have been offering a high-quality weatherization program. A number of them are extremely active, aggressive, and committed to conservation. Despite these efforts, initial public response was not significant. The Springfield Utility Board, after trying several promotional tools, began telephoning customers to obtain permission to conduct a home energy audit.

Response to the public utility programs was also poor because their space-heating customers have had little incentive to conserve electricity. As discussed earlier, these Oregonians have access to hydroelectricity generated by the Bonneville Power Administration. Currently, their rates are less than those of investor-owned utilities that have had to build expensive thermal generation facilities. As an example, a typical Port-lander served by an investor-owned utility paid an average of $37 for 1000 kWh of electricity in 1982, whereas the same amount of electricity was only $31 for a customer served by a publicly owned utility in Eugene (Public Utilities Commission, 1982).

As the rates charged to public utility customers continue to increase because of the legislated changes of the Northwest Power Planning Act and because of increased costs in electricity production in general, there will be a greater incentive for these customers to conserve. This situation applies to 20% of Oregon's electric residential consumers.

Administratively, the ODOE has experienced only a few problems with the part of the program dealing with publicly owned utilities. Since the changes in legislation in 1981, the oil heat weatherization program has also functioned smoothly. In 1982 there was little backlog in the energy audit for oil heat customers and relatively few complaints to the ODOE regarding the audits that had been carried out. During 1982, 5762 audits and 839 loans were made under the Oregon Oil Heat Program.

By the end of 1981 the public utilities and oil dealers had conducted 31,505 energy audits, and 1889 state-subsidized loans had been granted, for a total amount of $4,939,328. These numbers are relatively small compared to similar programs offered by the investor-owned utilities.

INVESTOR-OWNED UTILITIES SERVICES (HB2157)—1977

Description.   House Bill 2157 requires investor-owned gas and electric utilities to provide weatherization services to their residential space heating customers. When this bill was passed in 1977, Portland General Electric, Pacific Power and Light, Northwest Natural Gas, Cascade Natural Gas, California-Pacific Utilities, and Idaho Power were required to submit for approval by the Public Utilities Commissioner a residential energy conservation program that provides the following services:

Information, upon request, about weatherization and other means of saving energy.

Assistance and technical advice concerning advantages and disadvantages of methods of conserving energy, including an estimate of the cost to the customer of the weatherization services provided under the program.

Provision and installation of weatherization materials up to a value of $1500, except when storm windows are installed, in which case the maximum value is $2000.

Provision for customers to pay for the services over a reasonable period of time, not exceeding 10 years, and at a maximum rate approved by the Public Utilities Commissioner (this rate has been 6.5%).

Provision of utilities' own funds for loans to customers or arrangements for financing on behalf of the customers with lending institutions, in which case the utility guarantees the principal portion of the loan.

The utilities can be required to add to the customer's bill the cost of the weatherization provided. Renters are also eligible for these services but only with the consent of the owner of the dwelling who would be responsible for the costs. The general expenses of providing weatherization services can be recouped by the utilities through approved rate increases, while lending institutions can claim a tax credit for the difference between 6.5% and the going interest rate. This bill is administered by the Public Utilities Commissioner.

Response.   In 1978 the ODOE estimated that about 60,352 houses and apartments annually would be affected by the weatherization services program under House Bill 2157 and Senate Bill 371 (ODOE, 1978a, p. A18). Such participation would represent about 15% of the eligible housing stock. Under the assumption that 10% of residence owners annually would implement energy-saving practices not requiring a cash outlay and 5% would take additional moderate cost weatherization measures to retrofit their homes, a projected energy saving by 1980 of 6.1876×10^{12} Btu was made by the ODOE. It was assumed that houses that were weatherized to the existing building standards could reduce their energy consumption by as much as 40%.

TABLE 7.11. Response to the Weatherization Services Programs of the Six Investor-Owned Utilities in Oregon, from September, 1978 to December 31, 1981

Company statistics	
Total residential customers (1978)	978,641
Space-heating customers	392,839
Program statistics	
Total customers notified	933,061
Total customer inquiries	247,225
Home energy inspections	
Number requested	115,694
Number completed and pending	109,139
Weatherization jobs	
Number requested	46,444
Number completed and pending	44,414

Source: Public Utility Commissioner of Oregon.

Table 7.11 summarizes the response to the programs of the six investor-owned utilities under House Bill 2157 as of December 31, 1981. The effective dates of the program vary among the companies, making it impossible to calculate a total participation rate for the first year of operation. Portland General Electric Company's program is the oldest, commencing on May 20, 1978, whereas C.P. National Corporation has had the program with the shortest duration, with an effective date of October 19, 1978.

By December 31, 1981, approximately 153,000 homes had taken advantage of the programs under House Bill 2157, for a participation rate of about 8% (Public Utility Commissioner).

Evaluation.   While a 6.5% interest rate is permitted on weatherization loans under House Bill 2157, Pacific Power & Light created a program that offers an interest-free, deferred-payment loan.* The company finances the loans with its own funds and does not require their repayment until the home in question is sold. This program was adopted later by Portland General Electric and C.P. National and has been coined "operation insulation." These investor-owned utilities decided that energy obtained through the weatherization of their customers' houses is less expensive than energy acquired through new generation. All their customers benefit because rates do not increase as fast as they would if new electricity was derived from the construction of expensive generating facilities. These utilities are providing loans for all cost-effective

*A discussion of Pacific Power & Light's program is provided in Section 7.5.

weatherization, and they are insulating, at no charge, electric water heaters located in unheated areas.

The public response to the interest-free loan has been good and much better than the response to the 6.5% interest programs. By December 31, 1981, after approximately three and one-half years of operation, the electric utilities have made 30,769 loans at 0% and 1673 loans at 6.5%. The natural gas distributors, which had been offering loans at 6.5%, had processed 10,987 by the end of 1981. Since 23% of homes in Oregon are heated by natural gas, compared to 42% for electricity, the 6.5% program appears to have a lower penetration rate than the 0% program.

A later development has been the implementation of a cash rebate program by the investor-owned utilities as another option to the customer. According to the ODOE, the gas utilities are now permitted to offer loans at interest rates higher than 6.5% (between 6.5% and 12%) or a cash payment of 25% of the amount spent on weatherization up to $350. This new provision by the legislature was instituted recognizing that gas utilities have a different financial structure than electric utilities and that they do not receive the same benefits of avoiding high marginal costs through conservation.

The electric utilities may also offer a cash rebate option under a program coordinated by the Bonneville Power Administration. There are two options in the BPA program: a cash rebate or a zero interest loan. It is anticipated that this program will ultimately replace the various residential weatherization programs originated by the investor-owned utilities.

The interest-free loan is available to about 80% of all electric space heating customers in Oregon. Given the growth of demand for electricity relative to oil for residential heating, there is a potential for significant energy savings through this type of program. The six investor-owned utilities spent $82,381,662 for weatherization during the period between the effective dates of the loan programs and December 31, 1981 (Table 7.12). Over half of this expenditure was related to weatherization for windows.

Some problems accompanied the implementation of these incentives. At the outset, the Public Utility Commissioner wanted to avoid crash programs resulting in high demand concentrated over a short time interval. The commissioner's objective was to have utilities gear up for the demand and then maintain a steady staffing level with quality services. Northwest Natural Gas phased in its program by initially serving only its Portland customers. The utility eventually included its other six districts but not without some customer complaints about this delay.

The introduction of the interest-free loan programs was met with overwhelming demand. This demand resulted in a backlog of homes to be inspected and weatherized and in complaints about the length of wait for services. This high level of public interest had an effect on weatheriza-

TABLE 7.12. Direct Expenditure of Investor-Owned Utilities in Oregon Through Weatherization Loan Programs (from Program Effective Dates to December 31, 1981)

Ceiling	$11,938,442	15%
Floor	18,157,100	22
Window	45,277,603	55
Other	7,008,517	8
Total	$82,381,662	100%

Source: Public Utility Commissioner of Oregon.

tion contractors. They had to deal not only with peak demands but also with a temporarily dried up market when the customers of other energy suppliers held their weatherization plans in abeyance in the hope that their supplier would also offer interest-free loans. Other contractor complaints have concerned the bidding procedures used by regulated utilities and the burden of paperwork and time. The demand for most programs gradually tapered off, so that by late 1979 they began to operate more efficiently. Any criticism of these problems should be considered in terms of program uniqueness and sheer magnitude.

It is difficult to evaluate the success of loans offered by the investor-owned utilities. Participation rates are misleading unless more information is known about the number of homes already weatherized either because they are new or because of assistance provided by other programs. Clearly, however, House Bill 2157 has resulted in conservation that would not otherwise have occurred.

ENERGY SUPPLIERS' SERVICES (SB371)—1977

Description. Senate Bill 371 requires all energy suppliers producing, delivering, transmitting or furnishing heat, light, and power to provide energy conservation information services. This includes answering questions from the general public concerning energy conservation and energy-saving devices, providing inspections, and making suggestions concerning the construction and siting of both buildings and residences. The services provided must conform to rules prescribed by the Public Utility Commissioner.

SUMMARY OF ENERGY SUPPLIER CONSERVATION SERVICES PROGRAMS

Energy savings projected by the ODOE to result from the energy supplier conservation services programs were revised downward in 1979. In

1978 the ODOE had anticipated savings of 12.9×10^{12} Btu by 1980 (ODOE, 1978a, Appendix A). After analyzing the actual response to the various programs, this estimate was revised to 6.6284×10^{12} Btu (ODOE, 1979b, Appendix A). It is difficult to conclude that this level of energy saving is a success per se because of the absence of precedents or standards against which such an achievement can be measured. It is possible, however, to conclude that the benefits accruing to Oregon from the existence of these programs outweigh the costs, especially under circumstances where only cost-effective weatherization is performed. The fact that the investor-owned utilities have initiated interest-free loans without prodding by government suggests that their corporate objectives can be met while pursuing strategies consistent with the nation's common goal of energy conservation. Any incremental energy saving resulting from cost-effective weatherization is beneficial.

In 1979 the free home energy analysis and access to 6.5% financing was made available to qualified mobile homes, houseboats, and apartments through the passage of House Bill 2147. This development increased the potential energy savings of the program.

Elderly Low-Income Weatherization Refunds

Description. Senate Bill 4 (1977) made available to elderly low-income homeowners a weatherization cost refund of up to $300. In order to be eligible, the applicant had to be 60 years of age or older on January 1, 1977, and have applied for and received an owner property tax refund in 1977 based on 1976 income and taxes. In addition, the 1977 assessed value of the applicant's home had to be less than $30,000, the annual household income less than $7500, and the applicant could not be eligible for any federal weatherization program. Since federal programs have been aimed primarily at households with incomes below the poverty level, Senate Bill 4 was designed mainly for households with income levels between the poverty level (roughly $4000) and $7500. Weatherization reimbursements were restricted to one homestead only and were paid according to priority based on the date of the claim. Vouchers were to be honored by the Department of Revenue until the $4 milliion that was available in the state fund had been expended.

Response. The ODOE had anticipated that there would be 12,000 participants in the program in the 1977–1979 period. In fact, there were 4633, representing only 8% of those eligible (Department of Revenue, 1979). The program expired in June 1979, and the total response fell short of expectations. Estimated energy savings were only 0.1170×10^{12} Btu instead of the 0.4921×10^{12} Btu projected earlier (ODOE, 1979b, Appendix A).

Evaluation.   Although the bill performed a redistribution of income function, a primary objective was energy conservation. Between October 1977 and May 1979, $1,087,112 was distributed, for an average expense of $235 (Department of Revenue, 1979). The reason that this amount was less than the maximum available refund of $300 might be due to the "lumpiness" of weatherization expenditures and the fact that a relatively large outlay is burdensome to people who have fixed incomes.

When the Department of Revenue was implementing the program, information dissemination was a problem. Eligibility was determined by analyzing Oregon income tax returns. The department mailed a brochure to those who appeared eligible, informing them that they might qualify for the program. Unfortunately, many people in the target group did not have a high enough income to incur a tax liability and would therefore not have received a voucher through this system. To alleviate this problem, brochures were distributed to various places, such as senior citizens' clubs, where potential applicants could be reached. Another problem was that some people proceeded to weatherize their homes with materials that did not qualify as weatherization.

Senate Bill 4 complemented a similar program administered by the U.S. Department of Energy and implemented through various state agencies since 1976. The federal program has already weatherized thousands of homes in Oregon.

7.4   INVERTED RESIDENTIAL ELECTRIC RATES

Description.   Two investor-owned utilities in Oregon, Pacific Power & Light and Portland General Electric, have changed residential rate structures to encourage efficient use of the state's electrical generating capacity. The inverted structure charges varying rates, depending on the amount of energy the household uses. This type of rate structure rewards energy-efficient residential customers and discourages high consumption rates. PP&L has implemented a three-block rate structure, and PGE has put a four-block rate structure in place, as shown in Table 7.13.

Evaluation.   ODOE estimates that a considerable savings in electrical consumption can be generated by these structures. In the PP&L case, an overall change in residential consumption of 8.12% can be expected. For the PGE rate structure, a 5.75% change in residential consumption is predicted. Overall, a residential load savings of 4.72%, or 2.293×10^{12} Btu's, will result annually from the new pricing policy. This level of savings would be a remarkable result from a simple change in pricing policy.

**TABLE 7.13. Comparison of PP&L and PGE Inverted
Rate Structures**

	1st Block	2nd Block	3rd Block
PP&L[a]			
Consumption (in kWh)	0–300	300–600	600+
Rate (¢/kWh)	2.101	3.043	4.164

	1st Block	2nd Block	3rd Block	4th Block
PGE[a]				
Consumption (in kWh)	0–300	300–700	700–1300	1300+
Rate (¢/kWh)	2.623	3.161	3.988	4.775

Source: Oregon Department of Energy.
[a]Summer 1982 rates.

7.5 PACIFIC POWER & LIGHT COMPANY'S WEATHERIZATION PROGRAM

In addition to introducing a 6.5% loan as required by House Bill 2157, Pacific Power & Light Company (PP&L) implemented an interest-free, deferred-payment loan that became effective on August 1, 1978. The company saw a need for a cost-effective residential energy conservation program for the benefit of both their residential customers and the company itself. This need was perceived for several reasons:

Although PP&L promoted energy conservation, its customers did not perceive the company as really doing anything about it.

Despite the fact that acquiring energy through conservation is less expensive than producing it from new generating facilities, there was no "hard" conservation program.

From an analysis of consumer response to other federal and state programs, it was clear that the average consumer is not interested enough in cost effectiveness or the economics of energy supply to do something himself about weatherizing.

PP&L customers occupying a qualified single-family residence or duplex can request a home energy analysis to determine the cost effectiveness of installing additional weatherization materials (Table 7.14). Dwellings not served by the company prior to April 3, 1978 or that have

TABLE 7.14. Weatherization Materials Used by Pacific Power & Light When Cost Effective

1. Ceiling insulation up to an R-38 as well as, when deemed necessary, proper ventilation.
2. Floor insulation over unheated spaces up to an R-19 and ground covers in crawl spaces. Water pipes in unheated spaces will be wrapped as necessary to prevent freezing.
3. Storm doors and windows or double glazing of such, as well as weather stripping from caulking as deemed necessary.
4. Duct wrapping when associated with unheated space and forced air electric furnace.
5. Timed thermostats on forced air electrical systems when other weatherization services, as listed above, are also provided.
6. Water heater insulation blanket in an unheated space.

Source: Pacific Power & Light.

been converted to electric space heating subsequent to that date are not eligible for the program. With the consent of the customer, PP&L arranges and pays for all labor and material associated with installing cost-effective weatherization. The company also installs, free of charge, an insulation blanket on any electric water heater purchased on or before April 3, 1978, and located in an unheated space in the homeowner's building. The customer must repay the principal of the interest-free loan prior to or at such time as the ownership of the dwelling is transferred. As security for the loan, PP&L has a lien on the dwelling.

The cost effectiveness of the program was determined by comparing the cost of electricity generated by new plants with that acquired through weatherization. At the time of the program's implementation, the company determined that electricity from new generation would cost in excess of 4.2 cents per kWh. Based on an average investment of $1300 per dwelling, weatherization could provide electricity costing only 1.8 cents per kWh saved. The investment was treated just like that in new facilities and PP&L sought approval for regulatory rate increases to pay for the operation of the program and provide an appropriate rate of return.

All of the PP&L customers benefit from the weatherization activities of the participants through a slower growth in their level of energy bills. Those who partake are required to repay the loan to resolve the inequity to the nonparticipant who might have weatherized his home with his own funds prior to the beginning of the program. Another economic benefit is that rate payers financially support the company's investment

**TABLE 7.15.   Pacific Power & Light
Weatherization Program: Summary of Operations
as of December 1982**

Number of dwellings improved	19,698
Average cost per dwelling	$1,352
Annual kilowatt-hour savings	87,872,778
Average cost of weatherization material per kilowatt-hour saved	$0.0150
Administration costs per kilowatt-hour saved	$0.0034

Source: Pacific Power & Light.

in weatherization only until such time as the participants repay the loan. In the case of an investment in new generating facilities, rate payers must support them in perpetuity.

The energy "freed up" by the weatherization program is not enough to eliminate the need for new facilities. This energy will only temporarily offset the growth rate in electricity demand. PP&L officials had originally estimated that even if all of the eligible homes were weatherized (about 80,000 in 1978), the total number of kilowatt-hours made available would negate load growth for only two years. However, load growth forecasts have been revised downward more recently; so, conservation and other measures will negate load growth for longer periods than previously thought. From the company's point of view, the program still serves a useful purpose because the approval process for the construction of new plants is very lengthy, and the program helps the company to meet energy demands until such time as the new facilities come on stream.

Annual energy savings are estimated to be almost 5000 kWh per participant (Table 7.15). The average usage by PP&L's customers was 13,084 in 1978, so the saving per customer is on the order of 38% (PP&L, 1979a).

In August 1979, PP&L had weatherized 3310 dwellings for a participation rate among eligible customers of about 4%. This created a backlog in demand for PP&L's services that was not cleared up until late 1979. Despite the heavy initial response, PP&L had anticipated an even greater amount of interest. One explanation for this difference was customer suspicion about the terms of the loan. It appears that they were looking for "the catch." The benefits to the customer are truly great. Not only is all the work done for him, but also the loan is interest-free, the principal can be repaid with deflated dollars (as long as inflation continues), energy cost savings are reaped annually, and, when he sells his home, he can increase its price by the value of the weatherization to recoup the principal. Initially, news stories and a television advertisement provided sufficient publicity. An expenditure of about $65,000 was undertaken for the advertisement of the message, "Pacific pays the bill." Further promotion was not considered necessary until late 1979, when demand began to fall off.

Implementation of the program was a substantial undertaking and involved a lot of paperwork, filing, staff training, and use of computers. Administration expenses originally ran as high as 20% of the average cost of weatherization material per kilowatt-hour saved. The program is conducted through 17 district offices involving six administrators and about 150 energy consultants who perform the home energy analyses.

PP&L spent much of 1979 making the implementation system more efficient. One problem was delay in the disbursement of payments to contractors. Initially, paperwork was slow as contractor invoices were processed by hand in the district offices and then sent to the central office in Portland for final review. Now after some experience and adjustments, payment can be made within one month.

Under the original program formulation, three independent contractors were randomly selected by PP&L for each job and asked to submit a bid. In order to be eligible, a contractor had to post a $3000 performance bond and have $300,000 liability insurance, requirements that the contractors believed were too stringent. The bid with the lowest price was always chosen. If a customer wanted materials that were more expensive, he was asked to pay the difference. PP&L inspected the work done. PP&L has lately revised this procedure, making the homeowner responsible for finding a contractor to do the job to standard specifications.

Overall, the program has been quite successful. If future analysis indicates that actual energy consumption by PP&L's customers has indeed been offset by the weatherization activities, then the program will have been cost effective.

7.6 ENERGY CONSERVATION POLICY FOR PORTLAND (1979)

7.6.1 Portland Energy Conservation Demonstration Project

In August 1974, Neil Goldschmidt, the mayor of Portland, proposed to the U.S. Department of Housing and Urban Development (HUD) a two-year energy conservation demonstration project in his city. Its purpose was to analyze energy use in Portland and to develop a series of alternatives for conservation. This undertaking was intended to serve as a model so that the method of analysis and the alternatives could be transferred to other municipalities in the country. The demonstration project was completed in June 1977, and produced a comprehensive set of local-level energy conservation choices together with their costs, benefits, and impacts.

Included in the project was an analysis of energy conservation alternatives for the residential level. Forty conservation measures, including educational, mandatory, and incentive programs, were evaluated using computer-assisted engineering techniques to determine their cost and impact on energy savings. The study concluded that projected 1995 residential energy use in Portland could be reduced by 40% (Portland Bureau of Planning, 1977e). Four of the more effective conservation measures were as follows:

A requirement that homes meet weatherization standards prior to resale.

Provision of low-interest loans or tax credits to homeowners or builders for weatherization.

Federal standards for appliance efficiency.

Building codes with minimum insulation standards for new construction.

When the HUD project was completed, Portland proceeded to establish the mechanisms by which conservation measures could be chosen and implemented. An Energy Management Task Force was established to draft an energy conservation policy for city government. Some of the task force's recommendations were adopted in 1978. The HUD project determined that Portland could get by with 45% less energy consumption without affecting its mission. In order to achieve this reduction, the city has proceeded to implement all conservation investments that are cost effective.

The mayor appointed 15 citizens to an Energy Policy Steering Committee and 60 others to form a series of six technical advisory task forces. They reviewed the results of the HUD project and other information to formulate a policy that would increase the energy efficiency of existing structures and the transportation system of the city.

7.6.2   Energy Conservation Policy for Portland

Based on the recommendations of the Steering Committee, the city adopted a number of policies in August 1979. One of these makes the retrofit of buildings mandatory starting in 1984:

All buildings in the City shall be made as energy efficient as is economically possible as determined by costs of conservation action and price of energy. The retrofit of existing buildings for the purpose of energy conservation shall be accomplished through voluntary actions initially, with mandatory requirements imposed five years after the adoption of the Policy. Retrofit programs and the requirements must be cost-effective, comprehensive, and have the most equitable impact possible on all sectors of the community [Ordinance No. 148251].

Cost effectiveness is calculated using a 10-year, simple payback criterion.

Subsequent to the passing of the mandatory retrofit ordinance, voters in the city of Portland approved a referendum that requires that before the measure becomes effective in 1984 another referendum must again approve the measure. It is presently uncertain whether the measure will take effect in 1984.

The ordinance created the nonprofit Portland Energy Conservation, Inc. (PECI) to implement the policy. By establishing this organization the city avoids certain legal problems regarding acceptance of donations and relationships with contractors. Previously, when a contractor's work had not been up to standard, the city found it very difficult to remove him from an approved list because of the requirement for detailed documentation and justification.

The PECI was mandated to establish a "one-stop" energy conservation center. Its purpose is to facilitate voluntary compliance with the retrofit policy by providing a clearing house for information regarding all conservation programs available to Portland citizens. The PECI will also develop and implement a strategy to market energy conservation aggressively. It is anticipated that about $250,000 annually will be made available for this purpose.

In an effort to make available additional financial and tax incentives, the city will sponsor various bills for the consideration of federal and state legislatures. Where no other financing mechanisms for conservation are available, the PECI is mandated to establish a loan pool in cooperation with private lenders, as well as to facilitate the choice of financing options so that property owners can maximize financial benefits. The city considers the loan fund to be crucial to the success of the policy. Rather than place liens on houses, these loans will be guaranteed by the city. Portland found that its elder citizens were not participating in the interest-free, deferred loan programs of the investor-owned utilities because of the liens. Having lived through the Depression, these people have experienced loss of property through liens and therefore are particularly averse to them.

The eventual compliance with cost-effective retrofitting of residences will be enforced at the point of sale of the building and will include both owner-occupied and investor-owned properties. The city will rely on self-certification for the recording of retrofitted residences. This procedure is an attempt to avoid additional bureaucratic and administrative practices. Failure to submit a self-certification form will result in a "clouded title" on the person's property, preventing him from transferring title to the residence. In order to give the city some flexibility, each person will be permitted one clouded title. This clause will allow people, especially elder citizens who are short of money,* to sell their house without having to make a weatherization investment. The marketplace will help enforce weatherization because lenders will not mortgage a property that has a clouded title.

The retrofit policy is the first Oregon initiative to use a mandatory economic measure to encourage conservation at the residential level, and because of this fact, it has stimulated much debate and controversy. One element that will ensure the measure's acceptability, as well as its potential effectiveness, is the process by which it was formulated. Rather than imposing the policy on the citizens, the city identified key groups and influential leaders and asked them to serve on the Energy Policy Steering Committee. This committee held nearly 40 meetings with citizens and interest groups and over 1500 people participated in the review process. By having the committee set the objectives and formulate policies, the city is confident that the measures are acceptable to a broad base of citizens and that the public is more committed to achieving the goal of retrofitting 15% of Portland's housing units annually.

7.7 COMMERCIAL AND INDUSTRIAL ENERGY CONSERVATION PROGRAMS IN OREGON

Table 7.16 provides a summary listing of commercial and industrial energy conservation programs in Oregon.

7.7.1 Commercial Building Code

The Oregon Department of Commerce has responsibility for building code standards. In 1978 a revision of the building code was made to provide for more efficient use of energy for space heating and lighting. Limits were imposed on the amount of electrical power available for lighting according to the function of the lighted area. The Commercial Energy Building Code and Lighting Standards for Public Buildings specified certain measures to improve the efficiency of space heating systems in new buildings. These measures will reduce energy consumption in new commercial buildings by 2.2 trillion Btu's per year by the year 2001.

*About 50% of the homes in Portland are owned by persons over 60 years old.

TABLE 7.16.   Commercial and Industrial Energy Conservation Programs in Oregon

Program	State Bill Number	Estimated Energy Savings in 1981 (Btu's)
1. Commercial building code	n/a	0.426×10^{12}
2. State energy management program/ state buildings revolving fund	n/a	1.439×10^{12}
3. Institutional buildings grants program	n/a	1.605×10^{12}
4. Utility commercial audits	SB 111	n/a
5. Small-scale energy loans	SB 115	0.275×10^{12}
6. Business/industry tax credit	HB 2247	1.224×10^{12}
Total		4.969×10^{12}

Source: Oregon Department of Energy.
Note: n/a, not applicable.

7.7.2   State Energy Management Program and State Building Revolving Fund

Description.   The State Energy Management Program is a project administered by the Oregon Department of Energy to monitor and control the amount of energy used in state-owned and operated buildings. Established in 1976, the program had an objective of reducing this energy consumption in 1981 by 20%. The following measures were adopted according to the ODOE:

A computer record of energy consumption by agency, by building or facility, by year, by month and by quarter, and by fuel type.

Standard computer formats and procedures for the retrieval, analysis, and reporting of energy consumption data.

Routine quarterly reporting of monthly fuel use by agencies.

Individual energy use reduction goals for each building, typically a 20% reduction from 1976 use levels expressed as Btu per square foot of building space.

Requirements that a walk-through energy audit be conducted for all facilities of 20,000 square feet or larger.

An annual report of conservation results published by ODOE.

Mutual support among state agencies.

Help in qualifying for state and federal conservation project funding.

A State Building Energy Loan Fund with a $321,000 grant to finance installation of energy conservation devices in state buildings.

Evaluation. The SEMP objective of a 20% reduction in energy use has been met overall, but not on a building-by-building basis. The dollar value of energy saved since 1976 is over $18 million. It is difficult to determine costs of the project because these are charged to the individual departments involved and are not included in the annual SEMP report. However, most of the savings have come from improved building management practices and not from large capital investments, so that a large share of the $18 million reported in energy savings is likely to be net savings.

The total energy savings attributable to SEMP during 1977–1981 were 2.5×10^{12} Btu's. This amount is approximately 0.5% of Oregon's total energy use in 1977. Considering the magnitude of energy saved by the program and the net benefit realized by the state, SEMP must be considered a very effective program. It also offers the advantage of establishing an ongoing system to monitor energy use, which will realize energy savings in the future, as opposed to a crash program with only short-term benefits.

7.7.3 Institutional Buildings Grants Program

Description. This project is a U.S. DOE-initiated program that is administered in Oregon by the ODOE. The program provides an economic incentive for public institutions to take energy conservation measures. The federal government will provide funds for 50% of the cost of preliminary energy audits, detailed energy audits, technical studies, and installation of energy conservation measures. These measures must be cost effective according to a simple criterion of having a payback period of 1–15 years. The Bonneville Power Administration will also purchase from the state the annual electricity net savings resulting from the conservation measures in participating institutional buildings. BPA will pay the actual costs of the electricity conservation project or a rate of 29.2 cents per kilowatt-hour for the estimated amount of annual savings of electricity. This amount represents an estimate of the present value of BPA's avoided costs through electrical conservation measures.

This program has two phases: (1) preliminary energy audits and detailed energy audits and (2) technical assistance and conservation measures. The first phase funds a survey to identify low-cost conservation activities such as changes in operations and maintenance adjustments. The second phase finances detailed technical analysis of the feasibility of installing conservation and renewable energy devices as well as the costs involved in implementing these measures.

Evaluation. The IBP energy savings have been estimated to be approximately 0.33×10^{12} Btu's for 1982. It is predicted by the ODOE that more than half of the measures carried out would have been induced without IBP by energy price increases in the next 20 years. This suggests that the project is inducing significant measures that would not otherwise have been carried out and creating additional energy savings by causing some measures to be implemented earlier. The annual cost of the project, $500,000, is somewhat higher than other programs with a similar amount of energy saving in the commercial sector, such as SEMP. However, the net amount of energy saving is significant and, compared to residential programs, does a very effective job.

7.7.4 Utility Commercial Audits

Senate Bill 111 (1981) will require that electric and gas utilities offer energy audits to commercial customers in 1983. Many of Oregon's larger electric utilities already offer this service, and the city of Portland also provides commercial energy audits. ODOE estimates that as much as 1.718×10^{12} Btu's of energy could be saved annually by the year 2001 through this program.

7.7.5 Small-Scale Energy Loans

Description. The Small Scale Energy Loan Program (SELP) was started in 1980 as a change to the state constitution approved by referendum. It was amended in 1981 by Senate Bill 115. The program provides low-interest loans for individuals and businesses to implement renewable energy projects. The sale of tax-exempt bonds finances the SELP, and approximately $390 million remained available for loans on January 1, 1983. The interest rate available depends in part on prevailing market rates at the time of the most recent bond sale. In October 1982, a sale of $2.5 million in bonds allowed the SELP to offer funds at 11.9%. Unfortunately, this was also the market rate available in January 1983, when interest rates had declined substantially from levels of the previous October, thereby partly offsetting the advantage of SELP tax-exempt financing. On the other hand, in times of stable or rising interest rates, SELP funds would be a particularly attractive form of financing.

Evaluation. SELP had approved 48 applications for a total of $31,778,170 between March 1981 and December 31, 1982. These projects would produce 0.681×10^{12} Btu's of energy annually from renewable sources. Small hydro projects (less than 100 kW) and micro hydro projects (more than 100 kW) were the major uses for SELP funds in the first year. One problem for ODOE administrators is trying to channel funds into projects that are actually going to be constructed in the near term. Since each small hydro project approved typically costs several million dollars, allocating SELP funds to only one of these that does not get built can represent tying up a large portion of the annual budget unnecessarily.

The actual cost of SELP is difficult to determine. Because of the nature of financing, the cost is actually an indirect loss of state tax revenues. The interest in SELP financing has been good, with over 106 applications for loans by the end of 1982. Over half of these have not been approved. SELP will fund a significant number of renewable projects by the year 2001, with an annual contribution of 4.7×10^{12} Btu's. However, ODOE estimates that 70% of the financed projects would have been carried out without SELP. It is important that SELP administrators try to finance those projects that would not otherwise have gone ahead, and this will present a major challenge for the program.

7.7.6   Business/Industry Tax Credit

Description.   The Business/Industry Tax program was established by the state legislature in 1979 and significantly amended in 1981 by House Bill 2247. It was originally designed to provide incentives for industry to use waste heat and renewable resources. The amendments extended the program to cover conservation and recycling measures. Proposals must be precertified by the ODOE. The tax credit is 35% of the investment and must be claimed over a five-year period at a rate of 10% in each of the first two years, and 5% in each of the last three years. An annual ceiling on certification of $40 million of investment has been allocated to the program, but this total has not been reached since the program began.

Evaluation.   Table 7.17 shows a historical summary of the business tax credit program statistics.

The tax credit may be granted to energy projects with a variety of payback periods and different rates of return. Although ODOE tries to give preference to those marginal projects that would otherwise have been carried out with no tax credit, there are likely to be many viable projects that do get the tax credit. Thus, from the total gross annual energy savings of 1.2×10^{12} Btu's, a large fraction would likely have been implemented anyway. ODOE estimates that as much as 96% of the estimated energy saving would have been achieved without the tax credit. The net energy saving is only 0.054×10^{12} Btu's annually. In 1981 the ODOE precertified approximately $5 million worth of projects. The cost for these projects over a five-year period will be $1.75 million.

TABLE 7.17.   Historical Activity of the Oregon Business Tax Credit Program

Calendar Year	No. of Applications Received	Approved for Preliminary Certification	Total $ Prelim. Certified	No. Final Certified	Total $ Final Certified
1979	6	1	2,300	0	0
1980	62	44	16,260,000[a]	4	1,024,000
1981	118	98	4,850,000	42	2,077,000
1982[c]	104	82	16,090,000[b]	42	1,138,000
Total	290	225	37,202,300	88	4,239,000

Source: ODOE, 1983d.

[a]Includes a $12 million biomass facility.

[b]Includes a $10 million wind facility. In addition, an $8 million geothermal electric facility, a $10 million biomass facility and a $3 million wind facility likely will receive preliminary certification.

[c]Through September 15 only.

7.8   SUMMARY AND CONCLUSIONS

From the perspective of policy formulation, the effectiveness of a residential energy conservation program depends on two elements: the degree to which it is acceptable to the target market in terms of the ease of participation and the impact it has on curtailing energy consumption. The impact factor must be measured by the actual energy conservation induced by the program exclusive of that which would have occurred anyway.

Table 7.18 shows that the inverted rate structure for residential electricity consumption appears to be the most effective measure in terms of net savings. This estimate is based on ODOE assumptions about how consumers will react to the rate structures, and not on actual experience. These estimates may be subject to a high degree of error and must be viewed with caution.

The utility loans have been a very attractive inducement to conservation for residential consumers. The original estimates for energy savings by the ODOE have been reduced considerably, but the policy instrument is effective in terms of inducing weatherization by consumers who would not have otherwise done so. There are still some unanswered questions about how weatherization measures that reduce air loss from houses affect the indoor air environment. BPA has raised this issue and is particularly concerned about elevated levels of air contaminants that result.

Two other programs that have been moderately successful are the weatherization tax credit and the weatherization loan program for veterans. These have made relatively significant impacts on energy conservation mainly because of a good response rate arising from the ease of

TABLE 7.18. Summary of Residential Conservation Programs in Oregon

Program	Gross Savings	Net Savings	Net Savings as Percent of Total Energy Demand in Oregon (1981)
Inverted residential rates	2.293×10^{12}	2.293×10^{12}	0.4
Utility loans	0.859×10^{12}	0.453×10^{12}	0.08
Residential building code	0.407×10^{12}	0.407×10^{12}	0.07
State weatherization tax credit	0.894×10^{12}	0.267×10^{12}	0.05
Low-income grants	0.265×10^{12}	0.240×10^{12}	0.04
State veterans' loans	0.309×10^{12}	0.216×10^{12}	0.04
Pilot light ban	0.028×10^{12}	0.077×10^{12}	0.013
State renewables tax credit	0.037×10^{12}	0.019×10^{12}	0.003
Senior refund	0.017×10^{12}	0.015×10^{12}	0.002
State oil loans	0.028×10^{12}	0.011×10^{12}	0.002
Total	5.137×10^{12}	3.998×10^{12}	0.7

Source: Oregon Department of Energy.

participation. Since the tax credit was operationalized through the Oregon income tax form, it was a simple matter for many homeowners to understand its financial implications. The loan for veterans is packaged with a home purchase loan and involves no extra effort on the applicant's part. Although a veteran's loan for weatherization is properly classified as voluntary, the fact that the home must meet minimum weatherization standards in order for the veteran to be eligible for the purchase loan injects a mandatory element into the program. By packaging the weatherization loan with the home purchase loan, however, the weatherization requirement is easier to swallow. If eligibility for this program was not restricted to veterans, it would be even more effective.

The acceptability of the interest-free, deferred-payment programs of the investor-owned utilities is good, but their impact on energy consumption is limited because only space-heating customers of the utilities that provide the programs are eligible. Information on actual energy being saved is not available, but based on response rates, the interest-free loan is much more popular than the 6.5% loans being offered by both the publicly owned and investor-owned utilities. The popularity of the interest-free loans derives from the convenience of participating as well as the zero interest rate. The utility arranges all of the financing and weatherization installation.

The alternate energy device tax credits and loans have not made a significant impact on energy savings, although by triggering a new interest in solar, wind, and geothermal energy devices, they have good potential as policy instruments. The tax credit is more attractive than the loan,

**TABLE 7.19.   Summary of Commercial and Industrial Conservation
Programs in Oregon**

Program (1981)	Gross Savings	Net Savings	Net Savings as Percent of Total Energy Demand in Oregon (1981)
SEMP/State Building Revolving Fund	1.439×10^{12}	1.439×10^{12}	0.25
Institutional buildings grants program	1.605×10^{12}	0.990×10^{12}	0.17
Commercial building code	0.426×10^{12}	0.426×10^{12}	0.07
Business/industry tax credit	1.224×10^{12}	0.054×10^{12}	0.008
Small-scale energy loans	0.275×10^{12}		
Utility commercial audits	(starts in 1983)		
Total	4.969×10^{12}	2.909×10^{12}	0.5

Source: Oregon Department of Energy.

mainly because up to 65% of the cost of an energy device can be saved
by combining federal and state tax credits. Factors inhibiting program
effectiveness include high start-up costs and a need for a certain level of
mechanical knowledge on the part of the homeowner. It does appear that
these incentives have not significantly changed the structure of the solar
industry such that new and less expensive units have become available
as was hoped. The cost of solar devices remains so high that most would
not be economic in Oregon without the tax incentives.

The commercial and industrial programs in Oregon are much newer
for the most part and may be significantly revised once the ODOE gains
some experience in their implementation. The State Energy Management Program and the Institutional Buildings Program have achieved
notable levels of energy savings. If they are any indication of the potential savings available in the commercial/industrial sectors, then further
policy research should be directed to these areas. Policy formulation will
be more difficult than that for the residential sector, however, given the
diverse nature of commercial and industrial establishments.

Table 7.19 presents a summary of energy savings attributable to the
various commercial and industrial conservation initiatives. The estimates of net energy savings for some of the newer programs, such as the
Small Scale Energy Loan Program, may be significantly revised once
they have had more time to operate. The estimates of net savings for
SEMP and IBP are fairly accurate because they have been operating for
a number of years.

8

ENERGY CONSERVATION INCENTIVES: A WISCONSIN CASE STUDY

8.1 INTRODUCTION

Wisconsin is unique in the emphasis and scope of its efforts to limit energy consumption. It has moved primarily by using the power of the Public Service Commission, a utility-regulating body, to direct utilities to explore and, where feasible, implement rate structures consistent with energy conservation. The state has been in the forefront of electric tariff and natural gas rate design. Many of the programs discussed here have not been definitively evaluated because of their relatively recent imposition. In recognition of the tentative quality of proposed rate changes, Wisconsin's approach to energy conservation, while innovative, remains experimental.

8.2 THE STATE OF WISCONSIN: ENERGY ENVIRONMENT

Wisconsin is in the northcentral United States, is irregular in topography, and is characterized by sharp weather changes. Winters are cold,

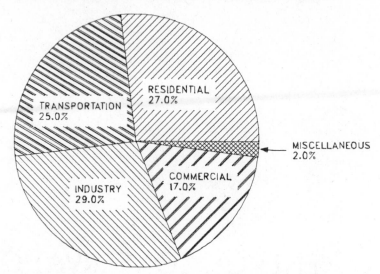

FIGURE 8.1. Categories of energy consumption in Wisconsin. (Source: Wisconsin DILHR, 1978d, p. 24.)

cloudy, and snowy while summers are warm with cool nights. The population is 66% urban and there has been a trend toward population migration from the state (Wisconsin, DILHR, 1978c). Figures 8.1 and 8.2 illustrate the sources and uses of energy consumed within the state. Residential heating uses 57% natural gas, 31% oil, and only 4% electricity. By 1985, oil use for heating is expected to increase to 39% while

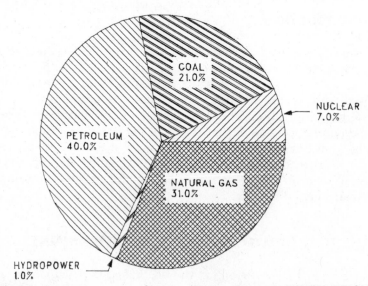

FIGURE 8.2. Wisconsin sources of energy. (Source: Wisconsin DILHR, 1978d, p. 24.)

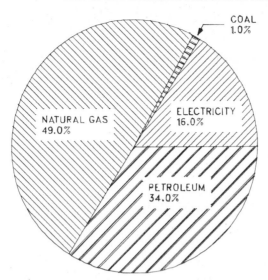

FIGURE 8.3. Energy types used in the Wisconsin residential energy sector. (Source: Wisconsin DILHR, 1978d, p. 28.)

natural gas will decline to 41%. The prospects of decreasing petroleum and natural gas supplies have increased the use of coal, particularly in electricity generation, and have caused state leaders to be extremely apprehensive about long-term energy supplies.

Figures 8.3 and 8.4 illustrate household energy use. It is apparent that home heating should be the focus of conservation efforts and, next to utility rate reform, Wisconsin has focused on the thermal integrity of its new and existing housing stock.

8.3 TIME-OF-DAY ELECTRICITY PRICING IN WISCONSIN

Wisconsin introduced a voluntary and experimental program of time-of-day (TOD, synonymous with time-of-use, TOU) pricing in 1975. By 1983 declining block rates had been eliminated and, although flat rates have now become standard practice, TOD rates have been made mandatory for large industrial and commercial customers since 1977. In fact, Wisconsin is one of the few states where the electricity energy charge is priced at the margin for peak and off-peak periods.

The history of time-of-use pricing involves a series of decisions by the Wisconsin Public Service Commission (PSC), which is charged with the power to supervise and regulate every public utility in the state. The PSC provides for a comprehensive classification of service that takes into account the quantity of energy used, the time when used, the purpose for which used, and any other considerations that may affect

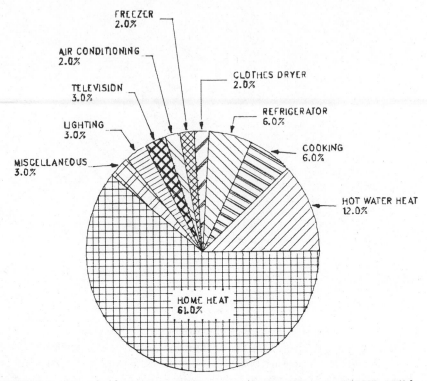

FIGURE 8.4. Household energy use in Wisconsin. (Source: Wisconsin DILHR, 1978d, p. 28.)

rates. All rates are subject to review and approval by the PSC. Decreases in charges may be made without a hearing, but any change potentially causing an increase in users' bills must be accompanied by an investigation and hearing by the Commission.

The rate reform hearing that triggered Wisconsin's involvement with time-of-use pricing was the case of *Madison Gas & Electric Co.*, 5 PUR 4th 28 (1974). In this hearing the commission indicated that rates should properly reflect the marginal cost of providing service to a customer as represented by long-run incremental cost. An investigation was undertaken into the possible benefits of a tariff system based on time-of-day metering with the cooperation of other utilities in the state. The PSC indicated that different rates should be developed for peak and off-peak usage and, pending the results of the investigation, ordered that a flat residential rate take effect with summer–winter differentials. (The new summer rate was approximately 50% higher than the winter rate and was in effect July through October.)

The rationale followed by the PSC in authorizing time-of-day rates was based on the cost of providing service and incorporated the following considerations (Malko et al., 1981, p. 159):

1. Class revenue levels were set at adjusted current revenue levels.
2. The off-peak energy rate was to reflect
    (a) System lambdas (a marginal cost measure of the next unit of production).
    (b) Average fuel cost.
    (c) Operating characteristics.
    (d) Voltage losses.
3. The on-peak energy rate was based on the cost differential indicated by the system lambdas.
4. The demand charge and the customer facilities charge were adjusted to equal the class revenue level and seasonal differential of the cost of providing service.

Generally, the PSC has been hesitant to impose rate increases on its residential customers.

The time-of-day rates approved by the PSC were challenged on environmental grounds by a group known as Wisconsin Environmental Decade, Inc. The group charged that the rates are promotional and encourage the use of electricity in the evening and winter when cheaper rates are in effect. They argued that the time-differentiated rates may actually increase the growth in demand for electricity and the group suggested that other techniques (namely, load management devices) are preferable environmentally. Previously, there had been a consensus among environmental groups in Wisconsin in favor of time-of-use pricing based on marginal costs, but the Environmental Decade broke ranks with the other groups. The thrust of their argument was as follows:

1. System load factor management should be the principal objective of pricing and supplying electricity.
2. Marginal cost cannot be used to promote economic efficiency because it does not hold true throughout the economy.
3. Time-of-use pricing will not save energy or capital, will not be fair, and will encourage base load nuclear technology.
4. The commission will not change time-of-use pricing once adopted because of inertia and political pressure.
5. Load control is superior because it brings certainty to reduced demand and eliminates the need to expand generation and transmission facilities.

The PCS responded that it found load control and time-of-use pricing to be complementary strategies and rejected the assumption that improvements in system load factor should be the primary objective for marketing electricity. The commission also found that time-of-use pricing would, in contrast to other rate reforms (including inverted or flat rates), stimulate and protect industrial and employment interests, reduce environmental damage, encourage conservation, and provide the consumer with freedom of choice.

Despite its interest in TOU pricing strategies to influence demand, the PSC has been involved, although to a lesser extent, in load management devices. On February 3, 1978, Wisconsin Electric Power Company requested PSC approval of a voluntary load management system for residential and farm water heaters, installation of a two-way communication system, and rate changes to permit a $1.50 per month customer incentive (credit) for volunteer participation in the load control program. The program was challenged by Wisconsin Telephone and Bell Laboratories, who charged that the load management communication system caused interference with telephone signals.

8.3.1 Commercial and Industrial Time-of-Use Pricing

Since January 1977, Wisconsin Power and Light has had a time-of-day rate in effect on a mandatory basis for about 140 large commercial and industrial customers. It was one of the first utilities in the United States to do so and the program has, by and large, been judged successful by both the utility and the affected customers. Time-of-day rates were implemented in December 1975 for customers with monthly demands greater than 1000 kW. The rates in effect now extend to customers with monthly demand exceeding 500 kW and, since 1978, a third rate has become operational for users in the 200–500 kW per month range as an option for smaller commercial and industrial customers who might profit from its adoption.

The aim of the utility was not only to improve control of its peak loads, but also to provide a stable rate that would not be continually revised to reflect increased generating costs. It was felt that industries could make future consumption and investment decisions more efficiently if the price of energy reflected the utilities' actual and projected costs. (Marginal costs are actually future costs, in contrast to the traditionally used embedded costs, which are based on costs that customer demand has caused in the past.) The rate structure adopted was based on three embedded and two long-run marginal cost studies commissioned by the utility. The on-peak energy charges were twice that of the off-peak charge.

Large commercial and industrial customers were chosen for three reasons. First, they already have the complex metering equipment in place; second, although they are few in number ($\frac{1}{10}$ of 1% of WP&L's customers), they are responsible for 25% of its total electric output; and, finally, they were thought the most able to adapt their operations in order to profit from off-peak operation.

Prior to undertaking the change in rates, WP&L conducted a comprehensive customer relations program initiating contacts with each affected customer. The company was very careful in explaining the rates to its customers, working with representatives of manufacturing and commercial organizations, and in graphically demonstrating to customers their own load curve characteristics.

Most of the changes in energy use patterns in response to TOD pricing occur with businesses and industries that have energy intensive processes, such as pumping, melting, drying, etc. These operations can be performed as easily during off-peak hours (10 P.M. to 8 A.M.). The shifting of personnel to night hours is more difficult as firms must weigh shift premiums paid to employees against potential energy savings. Some affected industries that have changed personnel to a night shift or added a third shift in response to TOD rates expect opposition from the unions when contracts are renegotiated. An alternative response on the part of some industries is the installation of holding furnaces, water storage units, and heat reclamation and refrigeration types of equipment to control demand.

In analyzing the data made available from the program, WP&L was not able to entirely quantify the extent of response to the time-of-use rates because of the numerous economic variables affecting energy consumption. These variables include the state of the economy, production demands, the weather, conservation efforts and mechanical or labor problems. Instead, it was decided to survey the customers directly. Eighty percent replied that the rate prompted them to analyze their operations in relation to the rate changes, and 40% stated that they had changed some operations in response to the rate. Companies that were able to do so shifted the work force to third (off-peak) shifts and some realized substantial savings. Generally, customers with traditional daytime or on-peak operations, such as department stores and schools, do not favor the rates; while those that are energy or labor intensive, particularly those able to employ three shifts, are in favor.

Overall, WP&L found that a minimum of approximately 25 MW, or 10% of its large commercial and industrial load, had moved off the system peak in one year (WP&L, 1978a). The cost savings to the utility with the reduction in peak period energy use are passed back to the customers through a fuel clause.

Although the rates were found beneficial for industrial and commercial customers, a comparison of costs and benefits to the utility for residential customers revealed that the rates were not cost effective. This was primarily due to the cost of the individual meters ($194), plus annualized maintenance and administrative costs. The necessary shift in demand required to break even was calculated as follows (WP&L, 1978a, Schedule 19):

Assumptions

Average residential coincident demand	1.25 kW
Average residential yearly usage	6000 kWh
Levelized carrying charge	16% (0.16)
Marginal demand cost of residential power	$189/(kW)(yr)
Difference between short-run on-peak and off-peak energy charge	$1.75/kW

Costs

Three-register meter (energy only)	$194/meter
Levelized annual cost	$194 × 0.16 = $31

Break-even

x = % shift in demand for break-even

$[(x)(1.25 \text{ kW})(\$189)] + [(x)(6000 \text{ kWh})(\$1.75)] = \$31$

$x = 9.1\%$

Therefore, a 9.1% shift in kilowatt-hours and kilowatts was deemed necessary to break even. "This means a shift of .11 kilowatts off the average residential coincident peak of 1.25 kilowatts and a shift of 546 kilowatt-hours per year to the off-peak period" (WP&L, 1978a, p. 62). It was concluded that a shift of this magnitude was not possible because most residential loads are thermostatically controlled and were difficult for residential customers to cycle without expensive timing equipment. These types of high demand loads included space heaters, water heaters, and air conditioners.

8.3.2 Residential Time-of-Use Pricing

Origin and Design of the Study

In 1974 the Public Service Commission ordered Madison Gas and Electric to submit a report indicating the feasibility and effect on its customers of various forms of time-differentiated and load-rate pricing, including interruptible service and time-of-day metering (Wisconsin PSC, 1977a, p. 1). Thereafter, this was expanded into a five-year federally assisted time-of-use pricing study involving approximately 700 specially metered customers and experimental rates. The goals were to determine the feasibility of residential time-differentiated rates and the impact on electrical demand (kW) and consumption (kWh) for a variety of rate structures ("time-of-day elasticity").

The study was designed and conducted by the utility, the commission, and officers of the Federal Energy Administration. It involved one year of design and preparation, three years during which the rates would be in effect, and a final year of analysis. The rates tested were as follows (Joiner and Ford, 1978, pp. 6–7):

1. Control rates: The standard "declining block" rate normally in effect. The first 200 kWh consumed in each month have one price, the next 1300 kWh have a lower price, and all kilowatt-hours above 1500 have an even lower price. Each customer also pays a monthly fixed fee that differs from urban to rural areas.

2. Flat rates: Each kilowatt-hour has the same price, which is the weighted average of all the utility's consumption in the above-200 kWh range from the control rate. The fixed charge is higher to cover the reduction in price in the below-200 kWh block.

3. Three-part rates: Has a fixed charge, a kilowatt-hour charge that is flat but lower than that of the flat rate, plus a *demand charge* based on the customer's highest instantaneous usage during a specified 9-hour period.

4. Time-of-day (TOD) rates: Has the same fixed charge as the flat rate, but the kilowatt-hour price is different during different parts of the day. All customers on a time-of-day rate have an on-peak period and an off-peak period. Weekends and holidays are considered off-peak.

The design of the experiment was constrained by the requirement that under law no more revenue on the average could be generated from these new rates than from the currently used declining block rates. In practice, this means that customers' bills were not allowed to vary by any significant amount (15% up or down maximum).

The experiment proceeded by way of a random sample, stratified by previous consumption, with higher levels of consumption more heavily sampled. Participation of the chosen customers was mandatory. In addition to the three rates tested, three different ratios of on- to off-peak period price were tested (2:1, 4:1, and 8:1), as well as three different durations of peak-hour periods (6 hours, 9 hours, and 12 hours). The summer and winter peak periods were as follows (Joiner and Ford, 1978, p. 7):

Summer

6-hour peak	9 A.M.–12 P.M. and 1–4 P.M.
9-hour peak	8 A.M.–5 P.M.
12-hour peak	8 A.M.–8 P.M.

Winter

6-hour peak	9 A.M.–12 P.M. and 5–8 P.M.
9-hour peak	8 A.M.–12 P.M. and 4–9 P.M.
12-hour peak	8 A.M.–8 P.M.

Appendix A to this chapter summarizes the experimental rates in effect.

In general, time-of-use rates, apart from other forms of marginal cost pricing or conservation-inducing tariffs, address the issue of deteriorating load factors (average load divided by peak load). The load factor problem is important not only because capital requirements are determined by peak loads, whereas revenues are derived from average loads, but also because it costs more to generate a kilowatt of electricity during the peak consumption portion of the day than off-peak. Originally, the commission turned to the evaluation of time-of-use rates because it recognized the difference in peak and off-peak operating costs. In addition, because construction and capital needs are related to peak usage, peak users are considered responsible for demand and capacity costs. A series of electric rate demonstration projects, sponsored by the Federal Energy Administration and reviewed elsewhere in this book, examined other innovations. Among those tested were experimental peak-load pricing (Wisconsin), inverted rate structures, rates approximating long-run incremental costs, various load control technologies (storage heaters and ripple systems), and conservation activities.

In addition to the Wisconsin project's stated goals of determining feasibility, effects, and demand elasticity, there were several secondary variables examined; for example, kilowatt-hours consumed in periods adjacent to peak periods, the trends in summer peak demand over time, changes in total consumption as a result of the rates, and differences in response among certain subpopulations.

Data Collection and Analysis

The first year of the project (1975) was spent collecting baseline data upon which the experimental rates would be founded. The energy use characteristics were modeled economically to predict the effect of rate modifications. One subgoal of the study was to determine the effect of time-of-use rates on the usage patterns of customers with various appliance combinations in order to predict the effect of such rates on system-wide appliance use patterns. Using this approach, the results of the experiment could theoretically be used to predict both how usage patterns would be affected by the imposition of a particular price ratio and how this effect might differ because of changing proportions of households with various appliance characteristics.

The models used to estimate the elasticity of substitution, in particular those pertaining to summer usage for WP&L (a summer-peaking utility), consisted of three separate formulations. First, the basic model permits elasticities to differ by month and by length of the peak period but does not control for the characteristics of various customers. Second, a modification of the basic model controls for various household characteristics, including the level of total kilowatt-hour consumption, appliance stock, and composition of household. Finally, the basic model is used to determine whether the elasticities differ between summer months or among the 6-, 9-, and 12-hour peak periods (Caves and Christensen, 1980, pp. 86–87).

Participating households were given supplemental information in their monthly bill indicating (1) the quantity and percentage of on- and off-peak energy used, (2) the quantity and percentage of on- and off-peak energy used during the previous month and year, and (3) the savings possible from a 5% shift from on- to off-peak usage. Compared with the cost of individual metering, which is very high, this form of feedback on energy consumption is a very effective and relatively costless form of incentive for conservation.

Three instruments were used to collect data: a specially designed meter, a mailed questionnaire to determine appliance stock and residence characteristics, and a second survey to measure attitudinal factors. The attitudes measured included those toward conservation, the environment, and perceived consequences resulting from the implementation of peak-period pricing.

The analysis of data proceeded in two phases. The first was to determine the dimensions of customer response to TOU rates, that is, the shift from peak to off-peak periods. Phase 2 concentrated on the nature of the shift—whether it was evenly distributed on weekdays, weekends, and holidays and how it varied seasonally. Phase 2 also evaluated the first phase, concluding that the appliance stock is an important determinant of customer response. Essentially, the degree of customers' response to the price structure is a reflection of whether or not they view peak and off-peak electricity as good or poor substitutes. If they are regarded as poor substitutes, even large changes in the price ratio of peak to off-peak usage will not induce customers to alter their established consumption patterns. Thus, the elasticity of substitution is the summary measure of the willingness of consumers to adjust their electricity consumption, and the initial stages of data evaluation focus on measuring that response.

Summary of Findings

ELASTICITY OF SUBSTITUTION

The measurement of elasticity of substitution (differing by month without reflecting the effect of different customer characteristics) is set forth

TABLE 8.1. Elasticity of Substitution for 6-, 9-, and 12-Hour Peak Periods, July and August

	6-Hour Peak, 9 A.M.–12 P.M. and 1–4 P.M.	9-Hour Peak, 8 A.M.–5 P.M.	12-Hour Peak, 8 A.M.–8 P.M.
July	0.133	0.108	0.091
August	0.143	0.119	0.112

Source: Caves and Christensen, 1980, p. 90. Reprinted with permission from *The Energy Journal*, Vol. 1, No. 2, © 1980 by the International Association of Energy Economists.

in Table 8.1. All the estimates are positive, which indicates that an increase in the peak to off-peak price ratio produces a decrease in the ratio of peak to off-peak consumption. The conclusion to be derived from these estimates is that time-of-use pricing "has a significant effect on the consumption patterns of residential customers: as the ratio of peak to off-peak price increases, the ratio of peak to off-peak consumption declines" (p. 90). Based on the elasticities contained in Table 8.1, the ratios of peak to off-peak consumption can be predicted for a variety of different price ratios. These are summarized in Table 8.2 for the month of August.

ELASTICITIES OF SUBSTITUTION FOR CUSTOMERS WITH DIFFERENT LEVELS OF CONSUMPTION

Households chosen for the experiment were initially divided into three use categories: low (100–500 kWh/month/1975), medium (600–1199 kWh/month/1975), and high (1200–1500 + kWh/month/1975). Customers

TABLE 8.2. Predicted Ratios of Peak to Off-Peak Consumption Resulting from Selected Peak to Off-Peak Price Ratios, August (Using Estimated Elasticities of Substitution from Table 8.1)

Price Ratio	6-Hour Peak	9-Hour Peak	12-Hour Peak
1:1	0.224	0.390	0.647
2:1	0.203	0.359	0.599
3:1	0.191	0.342	0.572
4:1	0.184	0.331	0.554
5:1	0.178	0.322	0.541
6:1	0.173	0.315	0.530
7:1	0.170	0.309	0.521
8:1	0.166	0.304	0.513
9:1	0.164	0.300	0.506
10:1	0.161	0.296	0.501

Source: Caves and Christensen, 1980, p. 92. Reprinted with permission from *The Energy Journal*, Vol. 1, No. 2, © 1980 by the International Association of Energy Economists.

TABLE 8.3.   Elasticities by Consumption Category

	July			August		
Consumption Category	6-Hour Peak	9-Hour Peak	12-Hour Peak	6-Hour Peak	9-Hour Peak	12-Hour Peak
Low	0.113	0.054	0.080	0.188	0.120	0.089
Medium	0.144	0.139	0.093	0.122	0.127	0.112
High	0.123	0.079	0.103	0.119	0.096	0.146

Source: Caves and Christensen, 1980, p. 94. Reprinted with permission from *The Energy Journal*, Vol. 1, No. 2, © 1980 by the International Association of Energy Economists.

were then randomly selected from each consumption range. High-consumption customers were oversampled because the cost associated with individual metering was thought to produce higher benefits when TOU pricing was used with higher-energy-consuming customers. Curiously, the hypothesis that the elasticities are the same for all three groups could not be rejected for any peak length or for either month (July or August). See Table 8.3. This means that elasticities are equal to those presented in Table 8.1 for all consumption groups. By implication, if the elasticity of substitution does not vary by level of consumption, the oversampling of high-consumption customers does not prevent the elasticities revealed in Table 8.1 from reflecting the average of the entire population of customers, rather than just the average of those customers involved in the experiment" (p. 93). Presumably, then, these results are transferable to populations outside the immediate experimental group.

ELASTICITIES OF SUBSTITUTION FOR CUSTOMERS WITH
DIFFERENT APPLIANCES

As mentioned, each participating household's stock of appliances was catalogued before the experiment to determine the effect of different appliance saturations on total consumption and the ability to shift usage. In order not to have to account for every type of appliance in use, the model's appliance variables were limited to air conditioners, dishwashers, electric water heaters, electric ranges, and electric clothes dryers. No estimates were made of the effects of these appliances individually, but together the five appliances significantly affected consumption patterns.

The effect of the appliances revealed itself in two ways. First, the presence or absence of the appliance affects the ratio of peak to off-peak consumption, and, second, the appliance may affect the ability of consumers to respond to TOU pricing. The first effect is the strongest in that appliance ownership increases the ratio of peak to off-peak consumption, even at the price ratio of 1:1 (Table 8.4). The elasticities of substitution shown in the lower part of Table 8.4 indicate that customers who own

TABLE 8.4.   Differences in Consumption Ratios and Elasticities of Substitution Due to Ownership of Major Appliances

| | Consumption Ratio at 1:1 Price Ratio | | | |
| | July 1977 | | August 1977 | |
Length of Peak	With Appliances	Without Appliances	With Appliances	Without Appliances
6	0.236	0.211	0.215	0.214
9	0.439	0.369	0.385	0.369
12	0.739	0.567	0.647	0.597

| | Elasticity of Consumption | | | |
| | July 1977 | | August 1977 | |
Length of Peak	With Appliances	Without Appliances	With Appliances	Without Appliances
6	0.193	0.079	0.211	0.104
9	0.200	0.030	0.187	0.062
12	0.143	0.034	0.143	0.071

Source: Caves and Christensen, 1980, p. 96. Reprinted with permission from *The Energy Journal*, Vol. 1, No. 2, © 1980 by the International Association of Energy Economists.

major appliances exhibit a higher elasticity of substitution between peak and off-peak periods. This finding leads to the following conclusions (Caves and Christensen, 1980, p. 95):

1. The residential consumption pattern is influenced by the ownership of appliances. Appliance ownership primarily affects the ratio of peak to off-peak consumption regardless of time-of-use pricing. . . .
2. The ownership of major appliances generally increases the ratio of peak to off-peak consumption.
3. In general, the ownership of appliances increases the response to time-of-use pricing.

The Wisconsin study concluded that similar stocks of appliances may result in highly dissimilar patterns of usage and that the more appliances a household possesses, the more able it is to respond to time-of-use pricing by using those appliances at different times. A household will benefit more from the institution of time-of-use pricing if, for example, it has a dryer whose use can be altered than if the household has no dryer at all; the household's potential for shifting is decreased by the lack of major appliances.

ELASTICITIES OF SUBSTITUTION FOR CUSTOMERS WITH DIFFERENT HOUSEHOLD CHARACTERISTICS

The variables tested here include the number of adults (age 18 or over), the number of children, the number of persons at home for different time periods during the day, the number of rooms per house, the education of the adults, and the urban–rural billing status. The findings were that the peak to off-peak consumption ratio does vary systematically with household characteristics, the most significant of which were the number of people in the household and the number of people at home between 6 A.M. and 6 P.M. Households with more people tend to have lower peak to off-peak consumption ratios, but households with more people at home during the day tend to have a higher peak to off-peak consumption ratio. Outside of these findings, there was no further evidence of any systematic relationship between any of the preceding household characteristics and the elasticity of substitution (Caves and Christensen, 1980, p. 97). The conclusion from these findings was that elasticity was the same across dissimilar households.

The final conclusion drawn from the preliminary evaluation of data obtained from the experiment was that Table 8.1, the elasticities of (summer) substitution, provided a reasonable estimate of the average population elasticity (if the related factors occur with equal frequency in the population and sample). Admittedly, the appliance mix in the sample differs from that of the population because of the stratified sampling procedure, but these effects could be corrected statistically. The elasticities in Table 8.1 were tested for statistical significance and were found to be representative for the two summer months tested. Insofar as time-of-use rates have been tested on summer peak load periods (the only period for which complete data have been analyzed), the researchers concluded that even with only a 2:1 price ratio, TOU pricing would result in a shift of kilowatt-hour usage from the peak to off-peak period. The average summer elasticity—the percentage decline in the peak to off-peak consumption ratio resulting from a 1% increase in the peak to off-peak price ratio—was found to be 0.117 (Caves and Christensen, 1980, p. 86). The study concludes that implementation of an 8:1 peak to off-peak price ratio would be expected to cause a 24% reduction in the peak to off-peak consumption ratio for the residential sector. It was shown that the test rates did reduce demand during a customer's peak period, but it remains unclear whether this demand was shifted to off-peak electricity consumption.

8.4   NATURAL GAS CONSERVATION INCENTIVES

8.4.1   Natural Gas Rate Design Proposals

As mentioned, Wisconsin has concentrated much of its conservation research on redesigning electricity and natural gas utility rates to fairly and accurately reflect the full social costs of energy consumption. Late in 1979 the Public Service Commission undertook a series of hearings to determine if, and in what manner, utility rates for natural gas might be redesigned to promote conservation. An overview of the testimony submitted to the commission illustrates both the positive and negative aspects associated with various rate design schemes, and which rate best serves as an incentive to conserve natural gas.

Behind each of the rate design proposals is the concept that the system effects of increased or decreased gas consumption should be reflected in the rate structure. As in the case of peak usage periods for electricity consumption, the costs of additional natural gas supplies to the system rise with increased usage. Using a principle of pricing resources at marginal social cost requires that these incremental supplies be priced at their expected cost over the utility's planning horizon, which is typically 8–10 years. The widespread practice of rolling in the cost of higher-priced supplemental natural gas supplies with lower-priced gas in order to average out costs to the end-user does not reflect the true cost of such supplies. If charging marginal cost caused revenues to exceed those prescribed, the excess could be rebated to customers through a lower initial block rate.

The commission was informed that the most important consideration in rate design was the price at which consumers would be induced to change fuels. It was suggested that as long as the end rate is less than the cost of the electricity, or fuel oil, alternative, massive switching will not occur. The danger is that the application of marginal resource pricing in the end blocks could result in customers switching fuels, resulting in unsold supplies and higher fixed costs spread over the remaining customers. This massive switching is referred to as the "domino principle" in the industry, and is the result of incorrect pricing. It is avoidable with correct tariffs. This problem occurs most often with industrial users, who are equipped to switch fuels at short notice. They leave the system as the higher-priced supplies are used and fixed costs are then spread over the remaining, residential customers, who cannot as readily leave the system. As remaining customers' costs increase, they too may finally leave the system and the domino principle continues. The conclusion is that pricing all gas at the margin could lead to an unstable market for natural gas (Wisconsin, PSC, 1979a).

One compromise suggested to the commission was an inverted rate with the tail block set at the cost per therm equivalent of the next best alternative source of energy supply. This would encourage more conservation and more gradual switching than either a straight marginal cost rate or a flat rate. If a customer is deciding to use more or less fuel, his decision will normally be based on the cost of his next best alternative or, in this case, the prevailing tail block rates. If he is deciding whether to remain a natural gas customer or switch fuels, his decision will be based on the average costs of natural gas consumption. Using an inverted rate that is designed to return the same revenues to the utility will produce the same average cost as a flat rate. Under the circumstances the average rate can be expected to be well below the end rate, which is equivalent to the cost of alternative fuels. Since an inverted rate would produce conservation efforts, more gas would be freed up, service would be curtailed less often, and the stability of industrial customers' supplies would induce them to remain within the system. The stability of industrial energy supplies was said to have had a favorable impact on employment levels as well. By contrast, it was argued that flat rates might not produce conservation behavior, and declining block rates would tend to encourage waste and retard conservation technology.

Because customers' bills could be expected to increase under any application of marginal cost pricing principles, it was proposed to the commission that the customer charge be abolished entirely, thus tying the bill more closely to actual gas usage.

When the commission addressed the issue of lifeline rates, it was argued that these types of rates would tend to discourage conservation by creating an artificially low price for natural gas. If lifeline rates were put into effect, the use of a means test to qualify for the rates would be necessary in order to avoid subsidizing low gas-consuming homeowners who have comfortable incomes. The standard criticism of lifeline rates was raised; that is, if the poor are to be given transfers, they should not be provided in the form of a subsidy on energy consumption. If subsidies are to be given, they should subsidize energy conservation, not energy consumption. Lifeline rates, it was argued, would lock the poor into an energy use lifeline cycle with no incentive to adopt conservation measures or reduce consumption. And further, the cost of the subsidy would be transferred via rate increases to commercial and industrial users. These users would, in turn, either pass on the higher energy costs to the public or be forced to lay off low-wage earners if the market for their products was reduced by the higher prices.

As of 1981, the PSC had prompted the use of a wide variety of experimental rates by the state's natural gas utilities. These included (Vierma and Malko, 1981, p. 7): two and three block inverted, flat, seasonal, rolling and fixed benchmarks, auction, and flat with optional auction. By September 1983, however, with the exception of a declining block rate created for Wisconsin Gas Company, all natural gas rates were flat. These rates decline between customer classes, but are flat within each class.

8.4.2 Natural Gas Energy Conservation Programs

Early in 1977 the Public Service Commission initiated hearings regarding a proposed order directing utilities to conduct energy conservation programs. Because the PSC is charged by law with protecting customer supplies of energy, it is particularly concerned with the supply of natural gas, which is finite and diminishing. This issue is distinct from supplies of electricity in Wisconsin, which potentially could be expanded. As illustrated in Figures 8.2 and 8.3, 31% of natural gas is utilized in the residential sector, while 49% of total residential energy consumption consists of natural gas.

The commission was additionally prompted to take action in the area of energy conservation by the federally enacted Emergency Natural Gas Act of 1977, which initiated a policy of interstate allocation in response to supply curtailments. A serious interruption was experienced during the winter of 1976–1977, when large commercial and industrial users had their supplies cut for the longest periods yet experienced. In an effort to prepare the state for any similar contingency, as well as adjust consumers to the projected declining supplies, the commission ordered gas utilities to submit gas requirements over a range of projected temperatures. Based on the sobering revelations contained in those reports, the PSC ordered the utilities to initiate programs for insulating the existing gas-heated residences in their service areas.

The utilities could choose among several alternatives: (1) utility loans for insulation installed by the customer; (2) utility financing and installation of the insulation for their customers; (3) utility provision and installation of insulation with the costs considered as a utility cost of service; or (4) any combination of the above or any variation. The utilities were directed to proceed by determining the number of insulated homes and the level of insulation and then calculating a target level of homes to be insulated prior to the forthcoming season.

Pursuant to the order, a variety of measures was submitted to the commission concerning the best levels of insulation, methods of financing, and alternative conservation measures deemed cost effective. Although the commission had specifically ordered that only ceiling insulation be installed, it was brought to its attention that ceiling insulation without provision for moisture control through ventilation and vapor barrier provisions was inadequate. Ventilation was therefore determined to be a necessary element in any insulation retrofit program.

Testimony was also submitted as to the inadequacy of ceiling insulation as the sole focus of a weatherization program. In particular, it was shown that other methods of energy conservation are much more cost effective than this type of insulation and that once homeowners had installed ceiling insulation, it might become difficult to convince them that further measures were also advisable. This led the commission to suggest that different measures be suggested to consumers along with information on their relative cost effectiveness; although ceiling insulation, weatherstripping and caulking, and automatic set-back thermostats were deemed to be the three most cost-effective conservation methods as well as those that yielded the quickest return per dollar invested.

The issue of what level of insulation would yield maximum energy savings was also addressed by expert testimony. The utilities themselves were cautious about recommending levels of insulation that would involve large sums invested without a correspondingly large reduction in utility bills. There was no agreement on an optimum level of insulation; rather that savings were most dramatic in a change from no insulation to R-19; that R-30 to R-35 provided the best level of insulation consistent with minimum cost effectiveness; and that standards above R-35 were not deemed worthwhile. In light of these findings the commission found that no single standard should be set, but to be eligible for financing, the minimum level of ceiling insulation should be R-25.

A comprehensive utility conservation program would involve provisions for consumer education and information, quality control and consumer protection, and financing. The utilities were each ordered to provide their customers with reliable information on the available types of insulation and the names of private and governmental agencies lending or granting money for installation. Public mass media campaigns were ordered in addition to information programs within the utilities' service areas. The cost of these education and information programs was to be treated as a utility cost of service. The programs were extended to include counseling at the individual homeowner level. The personal counseling services to be provided by the utility would include determining the kinds and amounts of energy conservation work required on the premises as well as subsequent quality control procedures to ensure that work was performed to standard.

The fear of poor quality materials and workmanship was frequently voiced, particularly in view of the increasing quantity of work to be performed after the commission's order. The utilities were therefore ordered to provide both a program of postinstallation inspection and a list of recommended applicators. The total program, therefore, would include a preinstallation inspection (energy audit), cost estimate and estimate of energy savings, and a postinstallation inspection certifying that the work performed was done correctly.

The financing schemes proposed by individual utilities differed widely in response to the commission's original order to provide some form of incentive to homeowners. Wisconsin Power and Light (WP&L) provided the most comprehensive program, including a gas conservation rate for customers having a residence that meets certain insulation standards, and other conservation measures at half price once the initial level of insulation was met. The program was estimated to cost over $8 million per year for 7 years, $5.6 million of which was allocated for interest expenses and losses to bad loans. Other utilities proposed reimbursement for all work once postinstallation inspection standards were met, with the programs proceeding on an area-by-area basis and costs translated ultimately into a surcharge on all retail gas rates. Two financial (mortgage) institutions submitted their own proposals as a special variety of home improvement loan for existing mortgages or an energy improvement option with new mortgages. In all cases, financing only without inspection was judged unsuitable, and the direct provision and installation of conservation measures by the utilities was considered anticompetitive and against public policy. Finally, in response to local groups' anxiety that they would be forced to pay for their neighbors who had not yet installed insulation, the commission ordered that utilities be lenders of last resort only and that as part of the information component of their conservation plan, they refer interested homeowners to financial institutions providing weatherization loans. Thus, virtually no obligation for financing conservation measures fell on the utilities.

Unique to the Wisconsin plan for a comprehensive insulation program was an attempt to provide for renters' rising energy costs. In recognition that it is seldom in the landowner's interest to weatherize his property when he is not responsible for utility bills, the commission cited numerous reasons why the state's interest in reducing tenants' high utility bills requires that landowners be forced to bring their property to certain minimum standards. In the absence of stronger, legislatively mandated powers to counter owners' low-maintenance actions, utilities were ordered to target areas containing rental units in which 80% or more were constructed before 1940 and indicate the extent of energy conservation modifications required. Owners would be required to meet the specifications or have gas service withheld. Owners were to be given 6 months in which to act. In December 1980 the PSC replaced its mandatory conservation standards for rental living units with a voluntary program. A prominent part of this voluntary approach was the use of utility sponsored life-style, energy audits for tenants (Wisconsin PSC, 1980 and 1982a, pp. 6–7).

Finally, to develop a comprehensive conservation plan, each utility was requested to devise a "conservation rate" to become effective for all customers participating in the weatherization program that would reflect savings to the system brought about by their actions. The rate would be based on the opportunity cost of alternative fuels. The utilities were ordered to submit their version of the rates relating to this specific class of customer, while the commission proceeded independently, as described above, in its attempts to develop an entirely new approach to gas pricing.

By 1983 the commission staff, after receiving the utilities' plans and comparing the various programs in effect, concluded that more efficient use of natural gas was in fact taking place. The drop in average residential gas consumption from 1974–1982 ranged from 29–34%, although the PSC concluded it was difficult to determine the role played by the conservation programs as distinct from energy price increases and other stimuli. The commission distributed a standardized reporting instrument in an effort to measure declines in consumption. As of 1980, over 290,000 energy audits had been completed, but only about 3000 postinstallation inspections had been requested or performed.

This is not to imply that no conservation measures have been undertaken. Although the utilities were not ordered to evaluate the success of their programs, each reported on the nature of its actions and assessed the program's success by a series of household surveys. By 1979 approximately 60% of the homeowners who had undergone a home energy audit, either personally conducted by utility representatives or by completion of a utility-provided home audit form (Appendix B to this chapter), had undertaken some form of conservation measures. Many customers reponded in particular to the Residential Payback Analysis, which quantified savings that could be expected from a variety of conservation measures. Wisconsin Natural Gas Company surveyed both homeowners who had undergone an audit and those who had not as a control group (Wisconsin PSC, 1979b, pp. 7–8). Of those who had *not* undergone a household inspection, 60.2% made conservation improvements; whereas for those that had the energy inspection the comparable figure was 62.7%. This preliminary study suggested that the program (at $62.26 per audit) did not warrant the savings, since conservation measures were undertaken as frequently in the absence of the audit program. Further, those measures that qualify as a conservation effort are often the most simple and least expensive; there were no consistent data on the types of conservation measures undertaken or savings by individual homeowners from which to gauge the effectiveness of the audit programs. This would seem to be a fault of the survey instrument rather than the audit program itself, however. Despite the results of the initial survey by Wisconsin Natural Gas Company, the PSC staff strongly supports the energy audit program, citing 90–95% approval by the almost 300,000 recipients.

A review of the activities undertaken in response to the commission's order to initiate a comprehensive conservation plan by Wisconsin Power and Light Company (submitted to the commission on September 28, 1979) indicated that preliminary activities involved efforts at public education and agency coordination. The utility conducted audits, and a comprehensive brochure and a dozen or more individual pamphlets on all aspects of home conservation measures are available upon request from the company. A study was undertaken to evaluate the effectiveness of WP&L's programs—in particular, to determine if audited customers were saving more energy than nonaudited customers, to ascertain the number of customers following suggestions, and to calculate the costs to the utility relative to the energy savings.

With reference to loan-financing and rate-making activities, the First Wisconsin National Bank initiated a specifically designed program of energy conservation homeowner loans, but this program was subsequently discontinued because of lack of public interest. Other statewide loan programs are available from the Farm Home Administration, and the Department of Housing and Urban Development as well as from Veterans' offices. Most of these loans are for low-income families. In general, the utilities have not been called upon to finance installation of conservation equipment but two companies offer loans at market rates.

The program to target and insulate rental properties has proceeded cautiously. WP&L established a number of rental advisory committees consisting of rental property owners, renters, realtors, bankers, and local government officials to examine ways of best developing projects fulfilling the commission's goal of weatherization. The first phase gathered data on the number and characteristics of rental properties within the various service areas and the energy use history of each property before proposing standards. In late 1979 WP&L conducted media campaigns promoting rental properties whose units met its own "WP&L Energy Conservation Seal" standards. In fact, the PSC decision to implement the voluntary conservation program was based on information provided to the commission by WP&L.

8.5   THE LOW INCOME WEATHERIZATION PROGRAM

In April 1982 the PSC moved to remedy an important deficiency in its energy conservation strategy. Wisconsin's eleven largest gas and electric utilities were instructed to use up to 0.2% of their gross operating revenue to assist low-income residential customers with energy conservation measures (Wisconsin PSC, 1982c). This "represented the first time the commission had required utilities to expend ratepayer money for accomplishing actual weatherization of selected customers' homes" (Wisconsin PSC, 1982b, p. 8). The specific instruments included low-interest or deferred-payment loans, and direct weatherization assistance. This action by the PSC was in response to belated recognition of the fact that low-income consumers in need of conservation equipment or materials are generally unable to repay any type of loan. Under the system currently in place, no installment payments are required and deferred payment loans may be forgiven after five years. While the program is aimed principally at homeowners, low-income tenants also receive direct assistance, and owners of rental property can benefit from loans with rates below the market.

8.6   ENERGY CONSERVATION BUILDING STANDARDS

The U.S. Energy Policy and Conservation Act (P.L. 94–163) gives the federal government the responsibility to foster and promote comprehensive energy conservation programs by establishing guidelines for state programs, providing coordination, and offering financial support. The respective states were to develop, by January 1978, mandatory thermal efficiency standards for all residences (Wisconsin, DILHR, 1978c, p. 8). In March 1978 the Wisconsin PSC held a series of hearings to request input to the state's Comprehensive Energy Conservation Plan, which consisted of requirements for new one- and two-family dwellings, as well as retrofit requirements for existing residences. The retrofit requirements were left to the individual utilities to develop and are discussed above; the new requirements were the object of further study to determine their effect on housing costs and population migration.

The state Department of Industry, Labor and Human Relations (DILHR) was concerned with the consumer economics of the proposed standards: in particular that (1) savings from energy conservation are not subject to income taxes and are therefore highly valued; (2) the value of savings increases at the rate of increases in the cost of fuel; (3) the money borrowed is at mortgage interest rates and the interest is deductible from income tax liability; and (4) the initial investment (in conservation improvements) is recoverable at the time of sale of the home.

The standards proposed to DILHR were studied in light of the trade-off between the effect on statewide housing costs and the estimated fuel savings that the standards would produce (purported to be equivalent to 600,000 barrels of oil per year). The standards themselves were performance standards rather than prescriptive measures; that is, they were not intended to discourage the use of any particular material or method of construction, but rather were thought to invite and stimulate imaginative approaches to building design.

The DILHR staff found that the proposed standards would tend to increase the initial purchase price of a house by $219–804; increase the down payment by $21–160; and, over the life of the mortgage, add between $564–2067 (at $9\frac{1}{4}\%$) to the cost of the home (Wisconsin, DILHR, 1978c, pp. 100–104). It was reported that these figures would tend to accelerate the trend for low-income people to be priced out of the new housing market. Evidence submitted from savings and loan institutions indicated that they projected a $4 billion decrease in accumulated savings in response to increased utility costs to consumers in the current year and these effects, too, would tighten the new housing money market. The final issue addressed was the possibility that dissimilar energy codes between states could result in significant population migration to milder climates in the West and South. Strict conservation standards were felt to preclude industry from moving into and expanding operations in a particular state, so both population and industry might leave the state, a trend already perceived by demographers in Wisconsin.

In light of these considerations, thermal efficiency standards were adopted that would reduce statewide residential space heating energy consumption by 55% by 1985 while adding an estimated 2% to the costs of new construction.

8.7   RENEWABLE ENERGY INCENTIVES

8.7.1   The Wisconsin Renewable Energy Program

The Wisconsin Alternate Energy Tax Credit was enacted on May 3, 1978. It offered tax credits to both individuals and corporations that installed qualified solar, wind or waste conversion energy systems. Eligibility for the tax credit was determined by compliance with performance criteria incorporated into the laws by reference to various existing nationally accepted standards. The credit applied to both existing systems that were modified and new systems. Manufacturers were given the option of applying for system or component approval by the Department of Industry, Labor, and Human Relations, which reduced the taxpayer's burden of proving eligibility of a given system. Total expenditure for the system had to be over $500 to be eligible and the percentage credit was to vary according to the year of installation, as follows (WP&L, 1978c). For an existing building:

Year Costs Incurred	Allowable Deduction (%)
1977–78	30
1979–80	24
1981–82	18
1983–84	12

The expenses covered included the design, construction, installation, and equipment of the system, including labor. The maximum allowable system cost eligible for an individual was $10,000 and no limit was stated for systems installed by business entities.

The act was extremely complex in two respects: first, it was ambiguous with regard to what systems qualified; and, second, it contained an information component requesting the taxpayer to use a formula to compare the present value of energy savings with the initial cost of the system. The suggested formula was optional, and other supporting information could be submitted demonstrating that the system was in fact cost effective over a 25-year period.

The formula was as follows (Wisconsin, DILHR, no date): In order to qualify for the tax credit,

$$FS \times IDF \text{ must be greater than } A + B,$$

where FS = annual fuel savings from the alternative energy system (the calculation to determine the annual fuel savings had to be submitted)

IDF = the inflation/discount factor = $\dfrac{1}{D-I}\left[1-\left(\dfrac{1+I}{1+D}\right)\right]^{N}$

D = discount rate
I = fuel inflation rate
N = design life of the system
A = total cost of the system minus state tax benefits
B = present cost of any estimated replacement costs of the alternative energy system (for which calculations had also to be submitted).

The act stated that in computing the energy savings information the following parameters should be considered: the average annual load, the percent of the load supplied by the alternative energy system, the design life of the system, the conventional and auxiliary energy costs, and the first costs of the system.

If the taxpayer could determine in advance whether the system complied with the act and was not deterred by the formidable calculations, he filed a separate schedule appended to his state income tax to claim the tax credit. All receipts had to be attached as well.

Within the first year of its inception, the Department of Industry, Labor, and Human Relations received approximately 285 applications for the tax benefits. Of the applications received, 178 were certified for the benefits and 25 declared ineligible. The proportion of ineligible systems that were home constructed was not tabulated. Most of the remaining applications lacked necessary supporting documentation.

The average system cost was $5000 and the average income tax benefit was $1225. No data were collected correlating the incomes of eligible recipients with the type or cost of individual systems. The applications received from businesses/corporations represented only 5% of the total, although the largest application was a waste-woodchip-fired boiler that provided space heat and hot water for a factory and office. This system cost $350,000 with a possible tax benefit of nearly $30,000.

The life cycle costing information requested of applicants did, however, provide a built-in method of evaluating the effectiveness of the program in terms of energy savings by requiring that each individual system be effective in its own right.

Following the first year of operation of the program, a number of significant changes were effected by the state government. First, on July 28, 1979, a direct refund program was instituted for individuals and businesses (see Appendix C). This removed some of the complexities associated with the income tax credit although a life cycle cost analysis is required for every system application. Second, on May 21, 1980, a property tax exemption was created for active solar and wind systems. Effective until 1995, this action was deemed necessary in order to abort the loss of savings due to increased property taxes which reflected the instal-

TABLE 8.5. Annual Applications and Certifications Under Wisconsin's Renewable Energy Program, 1979–1982

Year	Applications	Certifications
1979	412	333
1980	1078	614
1981	2759	1956
1982	3270	2944

Source: Bradt, 1983.

lation of renewable energy equipment. Third, the Governor's Budget Bill of 1982 reduced the payback period from 25 to 15 years; shortened the time period for submission of an application from 4 years to one; and eliminated waste conversion systems from the program. Fourth, as of January 1, 1984, a uniform 10% refund level was adopted for residences; eligible system costs were reduced from $10,000 to $7500 for individuals and from $1 million to $100,000 for corporations; and the eligibility criteria were altered to exclude any family with a gross income in excess of $40,000. This last measure was adopted in response to criticism that the program had been unduly benefitting individuals with higher incomes.

Tables 8.5, 8.6, and 8.7 summarize the program results for the period 1979 to 1982 inclusive. It is clear from the response and certification rates that this program, in combination with a federal 40% income tax

TABLE 8.6. Type of Applications Under Wisconsin's Renewable Energy Program, 1979–1982

Applications received		7519
Total certified		5849
Active solar systems		5166
space heating only	1973	
domestic water heating only	1820	
space and domestic water heating	1063	
swimming pool heating	298	
other	12	
Passive solar systems		569
direct gain configuration	266	
indirect gain configuration	297	
other	6	
Wind energy systems		103
Waste conversion systems (all corporate)		9
Ineligible applications		310
Applications still in process as of June 1983		1360

Source: Bradt, 1983.

TABLE 8.7.   Total System Costs for Wisconsin's Renewable Energy Program, 1979–1982

Year	Individuals and Business		Corporations and Cooperatives	
	Income Tax Credits	Direct Refunds	Income Tax Write Off	Direct Refunds
1979	$376,273	$ 16,805	$ 88,789	$ 0
1980	64,353	527,674	123,413	2,262
1981	3,708	1,800,122	0	74,515
1982	0	2,858,509	0	39,289
Totals	$444,334	$5,203,110	$212,202	$116,066

Source: Bradt, 1983.

credit (see Appendix C), has been quite successful in encouraging the adoption of renewable energy sources in Wisconsin. Of particular note are the estimated total energy savings associated with these financial incentives. Table 8.8 summarizes these data.

8.7.2   The Wisconsin Power & Light Solar Water Heating Program

WP&L initiated a modest program to install nine solar water-heating systems in 1978 and approximately 80 more in 1979. The utility chose two manufacturers through competitive bidding and trained four of its

TABLE 8.8.   Summary of Costs and Anticipated Benefits of Wisconsin's Renewable Energy Program, 1979–1982

Year	System Costs	Costs to Government	Expected Lifetime Benefits in Energy Saved	
			MMBTU	Dollars[a]
1979	$ 481,867	$ 2,936,449	2,050,850	$12,271,155
1980	717,702	4,644,629	1,205,297	13,908,840
1981	1,878,345	9,976,187	952,642	21,998,916
1982	2,897,798	14,932,432	908,548	23,299,029
Totals	$5,975,712	$32,489,697	5,117,337[b]	$71,477,940

Source: Bradt, 1983.

[a]"The conversion to dollars is based on the type and cost of the conventional fuel being used and the efficiency of the conventional energy equipment for each application and then totaled benefits in the form of energy savings" (Bradt, 1983, p. 6).

[b]This is equivalent to 1,185,802 barrels of fuel oil. This conversion is based on 5,754,000 BTU's per barrel and 75% combustion efficiency.

own installers. It arranged possible financing through outside lenders and guaranteed the installed units for one year. The Agreement for Purchase and Sale of Solar Water Heating System (Appendix D to this chapter) has no financing implications; it only permits continued monitoring of the system by the utility to determine its impact on energy consumption. The standard system provided by the utility costs $2995 installed and has qualified for the state tax credit. The system is designed for 40–60% of annual water-heating requirements and, although the utility allows a temporary, experimental solar space-heating rate, none is granted for water heating alone. Because the program is in its infancy and its goals are so modest, savings to the utility are not calculable; but savings to individual customers were estimated as follows (WP&L, 1978c):

| | Dollars Saved | | | |
	Electric	Gas	LP	Oil
1st year	$ 138	$ 62	$ 128	$ 132
5 years	810	364	751	774
10 years	1999	898	1854	1912
20 years	6315	2837	5858	6041

These estimates are based on a four-member family using 70 gallons of 140°F hot water per day, paying 4.08 cents per kilowatt-hour of electricity and per therm of natural gas, 50 cents per gallon of liquid propane, and 80 cents per gallon of oil, all reflecting an 8% increase in costs per year. These savings do not immediately make the proposed system cost effective in present value terms. The initial cash outlay of $2995 is offset at tax time by a federal tax credit of $799 and a state credit of $479, leaving the system cost at $1717. Most homeowners are unwilling to wait the necessary 10 years to recoup this investment and the response is understandably small.

8.8   CONCLUSION

Incentive measures to promote energy conservation in the state of Wisconsin have largely been directed at focusing consumers' attention on the reality of rising energy costs and diminishing supplies. Potential resistance to higher utility bills associated with many incentive measures was circumvented by experimental rate programs incorporating the requirement that there be no substantial change in participants' bills. The voluntary programs in time-of-use pricing and interruptible rates were put into effect without the requirement of a rate hearing. Such hearings are only required for proposed rate increases, and participation in a voluntary program would presumably not increase a customer's costs. The experimental nature of most measures reviewed here reflect the progressive, yet cautious attitude of the Public Service Commission in committing itself to any one conservation strategy. Court intervention initiated by affected parties, both environmental groups and selected industries, has raised issues that the commission must resolve. A parallel attempt in the state of New York to impose time-of-use rates on a group of approximately 175 customers who consumed large quantities of electricity was struck down by the state court, which ruled:

> No effort was made to attribute particular cost significance to the quantity of an individual's demand within those levels. As a result, if there is any reason to believe that a customer who consumes large amounts of electricity in a peak period is more responsible for the higher costs imposed on a utility at those times than another who receives less power in the same period, the present record does not supply a foundation for that proposition or an evaluation of its impact on rates. [*New York State Council of Retail Merchants, Inc. v. Public Service Commission*, 404 N.Y.S. 2d 899 (App. Div. 1978.]

Increasingly, large energy consumers are being asked or forced to face time-of-use rates or interruptible rates. If Wisconsin determines that it is in the public interest to initiate universal time-of-use rates despite the high metering costs, then this system can be compared with other tested methods of reducing energy consumption. Time-of-use rates are a means of legitimizing and facilitating the inevitable increase in energy costs that consumers must bear.

With regard to other Wisconsin conservation efforts, the statewide weatherization and building standards program has been comprehensive and commendable in its scope, particularly in its attempt to reach renters. Absent is a means of monitoring homeowners' conservation efforts—these measures could have been written into the initial order in hindsight. Where complex measures of effectiveness have been adopted, as in the solar tax credit, the product may have actually constituted a disincentive to homeowner participation in the program. Between these extremes some requirement is necessary to provide ongoing feedback to the government on the effectiveness of its actions other than an accounting of dollars spent.

The tentative nature of the programs results from the desire of the Public Service Commission to remain flexible with regard to its options while exploring potential programs thoroughly in advance of a decision. The range of voluntary programs is a propitious means of allowing consumers to choose those conservation methods best suited to their individual circumstances.

8.9 APPENDIXES

APPENDIX A. Time-of-Use Experimental Rates (1978)

Schedule Number	WPSC Rate Number	Number Hours On-Peak	On-Peak to Off-Peak Ratio	Urban or Rural	Monthly Fixed Charge	Winter (Nov.–June) Peak Hours (see Note 1)	Winter Energy charge per kWh (see Note 2) On-Peak	Off-Peak	Summer (July–Oct.) Peak Hours (see Note 1)	Summer Energy charge per kWh (see Note 2) On-Peak	Off-Peak
Rg-EU3	011	6	8:1	U	4.94	9 A.M.–12 P.M. and 5–8 P.M.	$.1065	$.0133	9 A.M.–12 P.M. and 1–4 P.M.	$.1266	$.0158
Rg-ER3	013	6	8:1	R	8.99	"	$.1063	$.0133	"	$.1258	$.0157
Rg-EU4	029	6	4:1	U	4.94	"	$.0795	$.0199	"	$.0940	$.0235
Rg-ER4	031	6	4:1	R	8.99	"	$.0794	$.0199	"	$.0937	$.0234
Rg-EU5	037	6	2:1	U	4.94	"	$.0528	$.0264	"	$.0620	$.0310
Rg-ER5	039	6	2:1	R	8.99	"	$.0528	$.0264	"	$.0621	$.0310
Rg-EU6	044	9	8:1	U	4.94	8 A.M.–12 P.M. and 4–9 P.M.	$.0835	$.0104	8 A.M.–5 P.M.	$.0977	$.0122
Rg-ER6	049	9	8:1	R	8.99	"	$.0830	$.0104	"	$.0970	$.0121
Rg-EU7	054	9	4:1	U	4.94	"	$.0676	$.0169	"	$.0791	$.0198
Rg-ER7	058	9	4:1	R	8.99	"	$.0673	$.0168	"	$.0788	$.0197
Rg-EU8	061	9	2:1	U	4.94	"	$.0490	$.0245	"	$.0572	$.0286
Rg-ER8	063	9	2:1	R	8.99	"	$.0489	$.0244	"	$.0573	$.0286
Rg-EU9	067	12	6.9/7.6:1	U	4.94	8 A.M.–8 P.M.	$.0688	$.0100	8 A.M.–8 P.M.	$.0761	$.0100
Rg-ER9	069	12	6.8/7.6:1	R	8.99	"	$.0684	$.0100	"	$.0764	$.0100
Rg-EU10	074	12	4:1	U	4.94	"	$.0601	$.0150	"	$.0664	$.0166
Rg-ER10	077	12	4:1	R	8.99	"	$.0599	$.0150	"	$.0667	$.0167
Rg-EU11	084	12	2:1	U	4.94	"	$.0462	$.0231	"	$.0524	$.0262
Rg-ER11	093	12	2:1	R	8.99	"	$.0461	$.0231	"	$.0526	$.0263

Source: Ray et al., 1978, Table 2. Reprinted with permission.

Note 1: Monday through Friday, except holidays.

Note 2: *Energy Cost Clause.* The adjustment shall consist of an adjustment to the on-peak and off-peak price per kWh based on a formula approved by the Public Service Commission. Ratio of on-peak to off-peak fuel adjustment per kWh will equal the on-peak ratio of the time-of-use rate of each customer.

APPENDIX B.   Home Energy Analysis

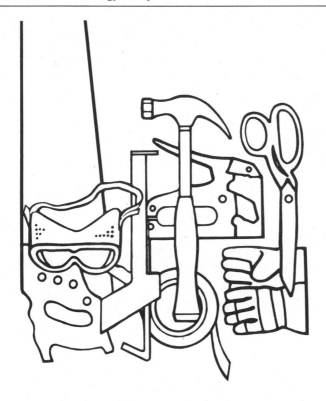

HOME ENERGY ANALYSIS

Wisconsin Power & Light Company

There are many things that can affect energy consumption such as life style, degree days, solar index, and wind. Because of these and other factors, all costs and savings presented on this form are estimates only and may vary greatly. The savings indicated may not be realized in your specific case.

Source: Wisconsin Power & Light Company. Reprinted with permission.

INFILTRATION outside air entering or filtering into the house through cracks or openings in the outside shell. Infiltration can account for 35 percent of heat loss in a typical Wisconsin home with reasonable insulation.

CAULKING (Refer to pamphlet or workbook page 18)

Caulking, (infiltration control) when done by the homeowner is an extremely low cost measure with high returns if properly done. Attention to detail is important, and the best available materials should be used.

☐ 1 Between window drip caps (tops of windows) and siding.

☐ 2 At joints between window frames and siding.

☐ 3 Where chimney, masonry, or other different materials meet the siding.

☐ 4 At joints between door frames and siding.

☐ 5 Where storm windows meet window frame except for drain holes at window sill.

☐ 6 Outside water faucets and other piping which penetrates the siding.

☐ 7 Dryer vent.

☐ 8 Dryer vent flap swings freely—lint buildup holding it open.

☐ 9 At sills where siding meets the foundation of the house.

☐ 10 At corners formed by siding.

☐ 11 Between porches and main body of the house.

☐ 12 Electrical entrance.

☐ 13 Phone and other wire entrances, e.g. cableTV .

☐ 14 Where the vent from the range hood passes through the wall, ceiling, or cupboard.

PROPER CAULKING METHODS

If there is too much of a gap to fill with a regular bead of caulk, you can use rope caulk or oakum instead.

Caulking will take a little practice. Make sure the bead overlaps both sides of the crack for a tight seal. The bead should be concave, if possible, and as deep as it is wide.

REMARKS: _____

WEATHERSTRIPPING

Weather stripping (infiltration control) also pays high returns, when done by the homeowner. Infiltration around windows and doors may account for 10-20 percent of total heat loss, with possibly half of this infiltration controllable by weather stripping. The best available materials should be used.

DOORS (Refer to workbook page 15)

☐ 15 Weatherstrip door(s)

Location

_____

☐ 16 Add a storm door(s) or insulated door or insulation equal to R-5 to

location _____ .

SWEEPS

INSIDE

DOOR SHOES

INSIDE

REMARKS: _____

WINDOWS

☐ **17** Repair broken/cracked pane in window.
Location

_____

_____

WINDOW JAMB → STRIP
CHANNEL
OUTSIDE
← SASH
a. OPEN
b. CLOSED
OUTSIDE
LOWER SASH BOTTOM RAIL
OUTSIDE

☐ **18** Add storms or double glazing to windows.
Location

_____

Approximate Savings $ _____ per year

Approximate Cost of Materials $ _____

☐ **19** Apply weatherstripping to windows. (Refer to workbook page 16)
Location

_____

This will return your investment in energy savings in approximately _____ years.

REMARKS: _____

BASEMENT (Refer to workbook page 11)

☐ **20** Insulate walls to minimum of R-5. (for heated basement)

☐ **21** Put storm windows on basement windows.

☐ **22** Insulate sill box area to R-19.

☐ **23** Insulate ceiling of basement to R-19. (For unheated basement)

☐ **24** Wrap heating ducts.

☐ **25** Insulate water heater and all exposed hot water pipes. Insulate cold inlet pipe 5 feet back from heater. Contact local office for instructions for insulating a gas water heater.

RIM OR BAND JOIST
SUBFLOOR
INSULATION
VAPOR BARRIER
TOP PLATE
SILL
VAPOR BARRIER
INSULATION FACE OF STUDS

☐ **26** Plug hole around soil pipe where it passes through the ceiling of the basement.

☐ **27** Add flow restrictors to faucets and shower heads.

☐ **28** Lower hot water temperature to 120° - 130°(140° if there is a dishwasher) by turning down the water heater thermostat.

REMARKS: _____

SPACE CONDITIONING EQUIPMENT (Heating & Cooling)

☐ **29** Change or clean furnace filters once a month during heating season.

☐ **30** Install an electric ignition on furnace.

☐ **31** Add fresh air intake to furnace.

☐ **32** Install automatic vent damper.

☐ **33** Put a thermostat on all heat tapes. Note: Most heat tapes do not have thermostats built in.

☐ **34** Clean or check air conditioner filter once a month during cooling season.

☐ **35** Install cover or remove room air conditioner from window at end of cooling season.

REMARKS: _____

CRAWL SPACE (Refer to workbook page 11)

- ☐ 36 Add _____ sq. ft. of ventilation.
- ☐ 37 Ground cover moisture barrier (6 MIL plastic).
- ☐ 38 Wrap heating ducts and water pipes.
- ☐ 39 Insulate ceiling to R-19.
- ☐ 40 Insulate walls to minimum R-5 and sill box to R-19.

SILL BOX   VAPOR BARRIER toward warmside
SPACE
INSULATION
SAND
VAPOR BARRIER

ATTIC (Refer to workbook pages 8, 9, 10)

- ☐ 41 Add _____ R-value of insulation.
- ☐ 42 Insulate attic access panel to R-30.

  Approximate Savings $ _____ per year

  Approximate Cost of materials $ _____

BAFFLE FOR VENTILATION
TOP PLATE

- ☐ 43 Clear insulation 3″ from sides of recessed lighting, and 24″ above.
- ☐ 44 Clear obstructions from vents.
- ☐ 45 Remove or slash vapor barrier on top of existing insulation.
- ☐ 46 Add _____ sq. ft. of high venting.
- ☐ 47 Add _____ sq. ft. of low venting.
- ☐ 48 Repair roof leaks.

3″
LIGHT FIXTURE

- ☐ 49 Vent bath/kitchen exhaust fans to the outside.

This will return your investment in energy savings in approximately _____ years.

REMARKS: _____

FIREPLACES

- ☐ 50 Add glass doors. On some fireplaces you cannot put on glass doors. Check with manufacturer before adding doors.
- ☐ 51 Keep damper tightly closed when not in use.
- ☐ 52 Add or repair damper.
- ☐ 53 Install fresh air for combustion.
- ☐ 54 Disconnect gas logs (fireplace).

REMARKS: _____

ADDITIONAL ENERGY SAVING IDEAS

- ☐ 55 Disconnect gas lights.
- ☐ 56 Fix leaky water faucets.
- ☐ 57 Disconnect gas pool heater.
- ☐ 58 Fastest way to save heating dollars is by lowering the thermostat, especially if left at lower temperature for 4 hours or longer. This can be done manually or with the addition of an automatic setback thermostat. To save cooling dollars in summer, use the air conditioner less or at a warmer thermostat setting.

Wisconsin Power & Light Company
HOME ENERGY ANALYSIS

Customer Name _____

Service Address _____

Account No. _____ Telephone _____

Date _____

Time _____

Age of Home _____

Log No. _____

Type of Home ☐ Own; ☐ Existing; ☐ Apt.-Units, No. of Units ____
☐ Rent; ☐ New; ☐ Duplex. ☐ Mobile;

	To Meet Standard ★	Suggestion

OUTSIDE: Sq. Ft. living area _____ Sq. Ft. of above ground wall _____ Sq. Ft. below Ground _____
☐ Type Wall Covering, ☐ Slate, ☐ Brick, ☐ Vinyl, ☐ Alum., ☐ Wood Other _____
Caulking: ☐ Windows; ☐ Corners; ☐ Doors; ☐ Sill; ☐ Faucet; ☐ Vents; ☐ Dryer Vents;
☐ Pipe Inlet; ☐ Wire Inlet; ☐ Brick Ledge/Chimney Siding meets Foundation

DOORS: Wood Sq. Ft. _____ , Storms _____ , Metal Sq. Ft. _____ , Insulated _____ ,
Weatherstripped _____ , Storms Weatherstripped _____

WINDOWS:
Single Pane Sq. Ft. _____ Double Glazed Sq. Ft. _____ Triple Glazed Sq. Ft. _____
Weatherstripped _____ % of Wall Area _____ Window Glazing _____

WALLS:
Present Type Insulation _____ Approx. R- _____ Vapor Barrier _____

BASEMENT:
Heated _____ ; Ceiling Insulated: R- _____ ; Walls Insulated: R- _____ ;
Sill Plate Insulated: _____ ; Sill Box Insulated R- _____ ; Floor Carpeted _____ ; Insulated R- _____

CRAWL SPACE:
Insulated R- _____ ; Vented _____ , Ground Cover Moisture Barrier _____ ,
Heated _____ , Heating Ducts Insulated _____ , Water Pipes Insulated _____

SPACE CONDITIONING:
Heating:
Brand: _____ Model No. _____ Type _____
Firing Rate: _____ Fresh Air Induced _____ No. Orifices _____ Filter Clean _____
Ceiling Cable _____ Heat Pump _____ Resistance Heaters _____ Heating Panels _____
Cooling:
Central _____ , Brand _____ , Model No. _____
Room Units _____ , BTU _____ , Watts _____ , (S)EER _____ , Filter Clean _____

APPLIANCES: Gas or Electric
Ranges ____ , Washer _____ , Dryer _____ , Microwave _____ , Dishwasher _____ , Fireplaces _____ ,
Refrigerator(s) _____ , Frost Free _____ , Freezer(s) _____ , Frost Free _____ , Water Heater _____ ,
Pool Heater _____ , Gas Lights _____ , Gas Grill _____ , Car Engine Heater (Elec) _____

ATTIC: Sq. Ft. _____ Sq. Ft. High Vents _____ Gable _____ Sq. Ft. Low Vents _____
Access Panel Insulated R- _____ , Recessed Lighting Covered _____ , Power Vent _____ , Vapor Barrier _____ ,
Vents Obstructed _____ , Evidence of Roof Leaks ____ , Floored Attic ____ , Exhaust Fans Vented to Attic ____ ,
Present Types of Insulation _____ , Depth _____ R- _____ , Depth _____ R- _____ , Total R- _____

FIREPLACE: No. _____ , Damper Closed _____ , Fresh Air Induced _____ , Glass Doors _____

ENERGY SAVING DEVICES:
Solar Heating _____ Recycler(s) _____ Wood Burners _____ Reclaimer(s) _____
Solar Assisted Water Heating _____ Flow Control Faucets _____ Auto Set Back Thermostat _____

OTHER: Leaky Faucets _____ Water Temp. _____ Dryer Vented Outside _____ Heat Tapes _____
Other _____

Customer _____
Signature

★ To qualify for an energy efficient home, these things must be completed.
WPL 4391-C

Auditor/Inspector _____
Signature

Approved _____

Source: Wisconsin Power & Light Company. Reprinted with permission.

◆▼ Wisconsin Power & Light

HOME ENERGY ANALYSIS REPORT

Prepared for _____ Phone _____

Address _____ City _____ Zip _____   PREMISE NO.

ENERGY CONSERVATION MEASURES

Analysis Number	DESCRIPTION OF MEASURES a	Recom-menda-tions Add	CONTRACTOR INSTALLATION b		DO IT YOURSELF INSTALLATION		ESTIMATED FIRST YEAR ENERGY SAVINGS	
			Estimated Cost c	Payback Years d	Estimated Cost c	Payback Years d	Dollars e	Units f
1)	Ceiling Insulation - Level 1 g	h	$		$		$	
2)	Ceiling Insulation - Level 2 g	R	$		$		$	
3)	Attic Ventilation		$		$			
4)	Masonry/Basement Wall Above Grade h	R	$		$		$	
5)	Frame Wall Insulation h	R	$		$		$	
6)	Basement Wall Insulation h	R	$		$		$	
7)	Sill Box Insulation h	R	$		$		$	
8)	Floor Insulation i	R	$		$		$	
9)	Storm or Thermal Windows ☐ Patio Doors ☐		$		$		$	
10)	Storm or Thermal Doors		$		$		$	
11)	Insulated Window Coverings/ Shaded Glass		$		$		$	
12)	Caulking Doors & Windows, etc.		$		$		$	
13)	Weatherstripping Doors & Windows		$		$		$	
14)	Water Heater Insulation Wrap		$		$		$	
15)	Clock Thermostat		$		$		$	
16)	Duct Insulation		$		$		$	
17)	Pipe Insulation		$		$		$	
18)	Flue Opening Water Heater ☐ Modification Furnace ☐		$				$	
19)	Elec. or Mech. Ignition Systems		$				$	
20)	Replacement Oil Burner ☐ j,k Replacement Furnace/Boiler ☐		$				$	
21)	Replacement Air Conditioner SEER		$				$	
22)	Elec. Load Management Devices l		$				$	

See reverse side for notes a through m.

TAX CREDIT CALCULATIONS

The federal government provides tax incentives to encourage the installation of energy conservation products. A tax credit applies to the cost of items installed after April 19, 1977 and before January 1, 1986. Details can be found in IRS publication 903, or by calling your local IRS office.

The sample calculation provided here applies to the following measures from above:

SAMPLE CALCULATION m	
Cost of measure(s)	$_____
Federal tax credits	$ -_____
Cost of measure(s) with tax credits deducted	$_____

The results of this Home Energy Analysis and the notes on the back of this form have been explained to me and I have been offered assistance in arranging installation and financing of the Energy Conservation Measures.

Customer Signature _____   Energy Consultant _____   Date _____

©1980, ENERCOM, INC. Tempe, AZ

WPL 4839-A   WHITE COPY — DISTRICT FILE   CANARY COPY — CONSUMER SERVICE DEPT.   PINK COPY — CUSTOMER

Source: Wisconsin Power & Light Company. Reprinted with permission.

◤◤ Wisconsin Power & Light

HOME ENERGY ANALYSIS REPORT

Prepared for _____ Phone _____

Address _____ City _____ Zip _____   Premise No.

RENEWABLE RESOURCE MEASURES

Analysis Number [ ]	CONTRACTOR INSTALLATION b		DO IT YOURSELF INSTALLATION		ESTIMATED FIRST YEAR ENERGY SAVINGS	
DESCRIPTION OF MEASURES a	Estimated Cost c	Payback Years d	Estimated Cost c	Payback Years d	Dollars e	Units f
* 1) Solar Domestic Water Heating	$		$		$	
2) Solar Replacement Swimming Pool Heating	$		$		$	
3) Active Solar Space Heating	$		$		$	
4) Passive Solar Heating - Direct g	$		$		$	
5) Passive Solar Heating - Indirect - Trombe wall g	$		$		$	
6) Passive Solar Heating - Indirect - Water wall g	$		$		$	
7) Active Solar Space and Water Heating	$		$		$	
8) Solaria/Sunspace Heating g	$		$		$	
9) Wind System h	$				$	

*Customer not interested in measures checked above.

TAX CREDIT AND REFUND CALCULATIONS

The federal government provides tax incentives to encourage the installation of wind power devices and active or certain passive solar heating and cooling systems. For precise information, contact the National Solar Heating & Cooling Information Center. Phone 1-800-523-2929 (toll free).

If a state credit is shown in the sample calculation, you may determine more specific information by contacting your local Department of Industry, Labor and Human Relations.

The sample calculation provided here applies to item(s) [   ] of the analysis.

See reverse side for notes a through j.

SAMPLE CALCULATION i	
Cost of System	$ _____
Federal Tax Credit	$ - _____
State Refund	$ - _____
Cost of System with credits deducted	$ _____

SOLAR/WIND SYSTEM FACTS

PASSIVE SOLAR				ACTIVE SOLAR				WIND	
System	Square Ft. Glazing	% Energy Supplied	Lbs. Storage j	System	Square Ft. Collector	% Energy Supplied	Lbs. Storage j	Suggested generator kilowatt rating for your home	KW
Direct				Active Solar Water Heater ☐Liquid ☐Air				Average wind speed at nearest wind measurement station	MPH
Indirect (Trombe Wall)				Active Solar Space Heating ☐Liquid ☐Air				Wind speed adjusted for your site	MPH
Indirect (Water Wall)				Active Solar Space & Water Heating ☐Liquid ☐Air				Based on adjusted wind speed, yearly elect. energy savings are approximately	%
Solaria/ Sunspace				Active Solar Replacement Swimming Pool Heating					

The results of this Home Energy Analysis and the notes on the back of this form have been explained to me and I have been offered assistance in arranging installation and financing of the Renewable Resource Measures.

Customer Signature _____   Energy Consultant _____   Date _____

WPL 4840   WHITE COPY — DISTRICT FILE   CANARY COPY — CONSUMER SERVICE DEPT.   PINK COPY — CUSTOMER

© 1980, ENERCOM, INC. Tempe, AZ

Source: Wisconsin Power & Light Company. Reprinted with permission.

APPENDIX C.   Wisconsin Schedule AE, Alternative Energy System Tax Benefit Claim

ATTACH PHOTO # 1 HERE
State of Wisconsin
Department of Industry,
Labor and Human Relations

**RENEWABLE ENERGY
DIRECT REFUND APPLICATION**

ATTACH PHOTO # 2 HERE
Safety & Buildings Division
201 E. Washington Ave.
P.O. Box 7969 Madison, Wi 53707

OFFICE USE ONLY

REFUND IS BASED ON TOTAL ELIGIBLE SYSTEM COST. ONLY those system costs adequately documented will be included in the computations.

ATTACH: 1. Photograph(s) of your system (Place applicants name on all photos).
2. Verification of costs (receipts, contracts, summary of eligible system costs).
3. Calculations which support renewable systems energy output.

RECEIPTS: 1. Number all receipts.
2. Receipts must include: Date purchased, to whom paid, to whom sold, what was purchased, cost of the purchased item, indication that item was paid in full, date of payment and who received payment.
3. If a single bill with multiple items, circle the claimed amount(s).
4. Itemize major costs in Section 13 on back of this form. If entire system has been provided by a single supplier, a signed contract marked paid in full and with date paid is sufficient receipt.
5. Cancelled checks are accepted as proof of payment, provided they are accompanied by receipts as described in #1 and #2 and are numbered to coincide with the appropriate receipt or bill.

1. COMPLETE ALL APPLICABLE ITEMS IF YOU ARE APPLYING FOR A DIRECT REFUND UNDER WIS. IND. 18.20

(Office Use Only) FILE NUMBER O					Name of Applicant (First, Last)	Designer, Dealer or Installer Name			
Street Address						Street Address			
City, State, Zip Code					Telephone	City, State, Zip			Telephone
Project Location (Street Address)					City, State	Zip	Code	County	Telephone
							(⋀ office use only)		

2. PROJECT DESCRIPTION (Include Photographs of System)

☐ Active Solar	1. ☐ Passive Solar	2. ☐ Earth Sheltered Passive Solar	3. ☐ Wind	☐ Photo Voltaic	7. ☐ Other (Specify)

3. RENEWABLE ENERGY SYSTEM USAGE (Check all that apply)

☐ Space Heating	☐ Space Cooling	☐ Domestic Hot Water	☐ Crop Drying	☐ Mechanical Power	☐ Other (Specify)

4. PROJECT CONSTRUCTION

☐ Contracted	☐ Home built (List components)	Manufacturer Name	**State of Wisconsin Material Approval Number:**
☐ Combined (Explain)		Model Designation	

5. SYSTEM TYPE (Check all that apply)

☐ Mobile Home	☐ Primary Residence	☐ Secondary Residence	☐ Agricultural	☐ Commercial	☐ Industrial	☐ Multi-Family	☐ Single Family
☐ Single Family-Attached duplex, condo, etc.	Other (Specify)						

6. FILL IN ALL BOXES THAT APPLY TO YOUR SYSTEM

Total cost of system (attach receipts/contracts): $	Annual building space heat/ space cooling load:	Have any Federal, State or Local grants or other public funds been used to construct this system? ☐ Yes ☐ No. If Yes. Amount $
Estimated first year savings: $	Number of persons in household:	Annual electrical load: (KWH/YR)
*Annual usable energy output:	*must be accompanied by engineering analysis unless information is currently on file with Renewable Energy section.	Back up fuel type:
Date of purchase:	Date installation complete:	Back up fuel unit price (e.g. cents per KWH):
Buildings first appeared on tax rolls: (2) ☐ Before (1) ☐ After April 20, 1977 (Required for application processing)	Is applicant owner of system?' ☐ Yes ☐ No	Owner of renewable energy system is owner of property on which its installed? ☐ Yes ☐ No	Owner is: ☐ Corporation ☐ Co-operative ☐ Individual (Includes all Others)

DILHR-SBD-5461 (R.11/82)

– SEE REVERSE SIDE –

7. ACTIVE SYSTEM INFORMATION

Collector Area (in Sq. Ft.-Total)	Orientation (Degrees from South)	Tilt Angle (Degrees from horizontal)

COLLECTOR TYPE (Check all that apply)	☐ Flat Plate 1.	☐ Flat Plate Tricke 2.	☐ Evacuated Tube 3.	☐ Concen-trating 4.	☐ Para-bolic 5.	☐ Photo Voltaic 6.	☐ Flat Plate Unglazed 7.	☐ Tracking 8.	☐ Other: 9.
COLLECTOR FLUID (Check all that apply)	☐ Air 1.	☐ Water 2.	☐ Water Anti-Freeze 3.	☐ Class I Fluids 4.	☐ Electric 6.	☐ Other:			5.

8. PASSIVE SYSTEM INFORMATION (Earth Sheltered Systems use SBD-5472)

TYPE OF SYSTEM (Check all that apply)	☐ Direct Gain (South Window)	☐ Thermal Storage (Trombe Wall)	☐ Solar Attic	☐ Greenhouse	☐ Envelope Design	☐ Skylights Atrilms, etc.

Sq. Ft. Area of South Facing Windows	Tilt of Windows (Degrees from vertical)	Type of Window Insulation R Value _____

Method of Positive heat distribution (if provided)	Type of Glazing: ☐ Glass ☐ Fiberglass ☐ Plastic ☐ Other:

9. WIND SYSTEMS

Tower Height and Manufacturer	Rated Output _____ KWH at wind speed of _____
Estimated average wind speed at site (ground level)	Is system interfaced with utility power source? ☐ Yes ☐ No

10. PHOTO VOLTAIC

Rated output/panel	Panel area
Total area of array	Is system interfaced with utility power source? ☐ Yes ☐ No

11. BATTERIES:

Are storage batteries utilized?

☐ Yes   ☐ No

If yes, a detail description of provisions taken to account for hydrogen produced during charging must be submitted.

12. STORAGE SYSTEM INFORMATION: (ACTIVE AND PASSIVE SYSTEMS)

Materials Used: Storage System Type and capacity (Complete all that apply)

Rock	Liquid	Battery	Concrete	Masonry	Phase Change	Other (Specify)
C.F.	Gal.	Amp. Hrs.	C.F.	C.F.	Gal.	
		Illuminated by the sun:	☐ Yes ☐ No	☐ Yes ☐ No	☐ Yes ☐ No	☐ Yes ☐ No (only if applies)

13. ITEMIZE MAJOR COSTS (See instruction #4 on front) (Attach additional sheets if necessary – use the same format as below)

Material Description	Quantity	Where Located in System	Cost
1.			$
2.			$
3.			$
4.			$
5.			$
6.			$
7.			$
8.			$
9.			$
10.			$
11.			$
12.			$
13.			$
14			$
15.			$
16.			$
		TOTAL	$

14. CERTIFICATION:

I certify that the information contained in this application is accurate to the best of my knowledge.

Applicant Signature and Date

Federal Tax Credit

Conditions:

1. The equipment must be installed on your
    principal residence (not on a vacation
    home). You cannot claim a credit for
    the expenditures until installation is
    completed, or, in the case of a
    dwelling being constructed or recon-
    structed, until you begin to use the
    home.

2. You must be the original user of the
    equipment claimed for credit.

3. The equipment must be expected to
    remain in operation for at least five
    years.

4. Only expenditures between January 1,
    1980, and December 31, 1985 may be
    claimed.

5. If the government establishes any
    standards for system quality or
    operation, your equipment will have to
    meet the standards. (No standards have
    been issued as of September, 1982).

The tax credit is 40% of the first $10,000
of qualifying expenditures. Such
expenditures include labor costs,
installation, and equipment, but exclude
structural components of the dwelling.
(Exception: solar panels that are a part of
a roof qualify). You may claim the credit
more than once if you expand the existing
system or purchase more than one system,
but the maximum cumulative credit is $4,000
on one residence.

To claim the tax credit, you must file
Form 5695 with your income tax return.
After calculating the amount of credit you
qualify for, subtract this credit from your
tax liability. You do not need to itemize
deductions to claim this credit. If you
itemize deductions, do not list your
equipment expense or your tax credit as an
itemized deduction. If your tax credit
exceeds your tax liability, you may carry
over the unused amount to a future tax
year. More details on this are available
from the Internal Revenue Service in
publication 903, "Energy Credits for
Individuals". For more information, call
the Internal Revenue Service toll-free
number, 1-800-452-9100.

RENEWABLE ENERGY

State Direct Refund

Conditions:

Expenses for the design, construction, equipment, and installation of a renewable energy system qualify. Passive solar installations only qualify under special circumstances. All equipment must meet performance and durability standards of the Wisconsin Administrative Code, Chapter IND 18 as approved by the Wisconsin Department of Industry, Labor and Human Relations (DILHR). Also, the system must pay for itself in not less than 4 years and not more than 15 years.

1. Expenses must be incurred between January 1, 1979, and December 31, 1985.
2. The cost of the system must exceed $500 in a single year.
3. Once a refund is claimed, subsequent owners of the system may not claim a refund for the same system.

The amount of refund you receive is calculated on the first $10,000 of eligible costs of your system. The refund levels as percentages of system costs, are as follows:

Year Costs Were Incurred	Old Buildings	New Buildings
1977-1978	30%	20%
1979-1981	24%	16%
1982-1983	18%	12%
1984-1985	12%	8%

A renewable energy refund packet is available for $2.10 (prepaid) including sales tax and postage from:

> Department of Administration
> Document Sales Distribution
> 202 South Thornton Avenue
> Madison, WI 53702

If you need further information on this refund, contact:

> Wisconsin Dept. of Industry, Labor and Human Relations
> Safety in Buildings, Room 101
> 201 E. Washington Ave.
> Madison, WI 53702   266-1149

Effective January 1, 1981, wind and active solar energy systems are exempt from local property taxes, provided the system meets DILHR standards. This property tax exemption is effective through December 31, 1995.

 Wisconsin Division of State Energy

Department of Administration
101 South Webster Street
P.O. Box 7868
Madison, Wisconsin 53707 (608) 266-8234

APPENDIX D.   Wisconsin Power & Light Company, Agreement for Purchase and Sale of Solar Water Heating System

Wisconsin Power & Light Company

Investor-owned Energy

222 West Washington Avenue   P. O. Box 192   Madison, Wisconsin 53701   Phone 608/252-3311

AGREEMENT FOR PURCHASE AND SALE OF

SOLAR WATER HEATING SYSTEM

In consideration of the sum of \$_____ of which a minimum
of ten percent (10%) is paid herewith to Wisconsin Power and Light
Company (hereinafter WP&L) and the balance to be paid within thirty (30)
days of the installation of the solar water heating system, _____

_____ agree(s) to purchase and WP&L agrees to
sell one solar water heating system subject to the following:

1.   The purchase price includes the cost of installation which will
be performed by WP&L or its agent.

2.   After installation of the solar water heating system, all care
and maintenance of the system is the duty and responsibility of the
purchaser.

3.   WP&L guarantees for a period of one year from the date
installation is commenced the satisfactory installation of the solar
water heating system.   Any other warranties pertaining to the solar water
heating system or its component parts are those made by the manufacturer.
WP&L expressly disclaims all other warranties, either expressed or
implied, including any implied warranty of merchantibility or fitness for
a particular purpose, and neither assumes nor authorizes any other person
to assume for it any liability in connection with the sale of the solar
water heating system or its component parts.

4.   The purchaser agrees that any sale or lease of the home will
reference this agreement, and that any sale or lease will be made subject
to the terms and provisions of this agreement and that all such terms
will be binding upon successive owners and tenants of the home.

5.   Delivery and installation of the solar water heating system is anticipated to begin on or about _____ and will require up to three weeks.

6.   Upon delivery and installation, title to the solar water heating system vests in the purchaser.

Dated this _____ day of _____, 19____ .

WISCONSIN POWER AND LIGHT COMPANY

By:_____
Consumer Services Representative

_____
Manager

Dated this _____ day of _____, 19____ .

_____
Purchaser

_____
Purchaser

_____
Street Address

_____
City, State, Zip

F57:13:03
2-12-80

Source: Wisconsin Power & Light Company. Reprinted with permission.

9

SUMMARY AND
CONCLUSIONS

This book has provided a detailed examination of several subjects critical to the formation and execution of energy conservation policy: pricing issues, case studies and evaluations of existing programs, issues of program transferability, and the use of econometric data.

This chapter summarizes and highlights a number of central issues in this study: (1) the role of econometric data in energy policy analysis, (2) essential elements in program design and implementation, and (3) the need for the collection and evaluation of behavioral data associated with energy consumption by consumers.

9.1 THE ROLE OF ECONOMETRIC DATA

Econometric estimates of price elasticities of demand for fuel are indispensable ingredients in the analysis of energy use and the formation of energy policy. There are several problems in using these data, however, that necessitate the utilization of complementary types of analyses such as case studies and behavioral questionnaires. In particular, two problems can be identified that constrain the usefulness of econometric analysis as a sole guide to energy policy formation. First, econometric data in the field of energy use are frequently too aggregated for the derivation of precise and robust conclusions. This shortcoming can be a function of either experimental design or data limitations and constraints. Second, the econometric data are generally out of date. This is particularly important in light of recent major dislocations in energy price and supply, and the change from a period of declining real energy prices to a period of sporadic increases in the real cost of fuels.

In contrast to conventional microeconomic analysis of the firm, the energy policymaker is less concerned about distinctions between "elastic" and "inelastic" price responses. Even though it is clearly desirable for a policymaker to be faced with an elastic response, potentially large absolute energy savings can still be achieved with relatively inelastic responses. This is particularly important where the energy form in question has widespread and heavy use. Nevertheless, the direct economic burden on the consumer is greater with inelastic than with elastic demand.

The econometric data summarized in this book support the intuitive conclusion that policymakers generally have less flexibility in the short run than in the long run. Energy use in the short term is tied to current capital stocks and behavioral patterns that tend to display substantial inertia. This social resistance to change is partly due to previous conditioning from low energy prices, increasing incomes, and the general social phenomenon of rising expectations.

Finally, the econometric data tend to illustrate the fact that there is more "slack" in the use of some fuels than in others. For example, gasoline use can be cut back by using alternative transportation forms (where these modes are available) and by reducing pleasure driving and other nonessential travel. In contrast, residential electricity use might not be as easily curtailed, especially if it is tied to lighting, cooking, heating, and other essential activities.

9.2   ESSENTIAL ELEMENTS IN PROGRAM DESIGN AND EXECUTION

9.2.1   Successful Policy Elements

We have identified at least nine elements that are considered to be essential in the successful implementation of programs of economic incentives for the conservation of energy in the consumer sector. Brief examples drawn from the case studies are provided for purposes of illustration.

Ease of Participation

The prospects for program success are improved through the use of in-place and broad-based administrative systems. These include the use of tax incentives that are incorporated in annual income tax forms or special weatherization loans tied to current home loan programs. In addition, nongovernment initiatives such as installation of weatherization by utilities can increase the acceptability and coverage of energy conservation activities.

Significant Monetary Incentives

Particularly effective are low (6%) or zero interest loans (as offered in Oregon), large tax credits (55%), and deferred paybacks (e.g., until resale of home).

Extensive Information Diffusion

Oregon's weatherization tax credit was helped immeasurably by a heavy public information campaign in all the major media. The importance of this factor cannot be overestimated. Potentially useful and effective programs of economic incentives can be rendered ineffective because of a lack of consumer awareness. This fact highlights a key conclusion in the area of energy policy: Individual instruments, such as economic incentives, cannot be employed in isolation. They must be part of an integrated package of policy tools that includes information and education programs and mandatory regulations.

Consultation and Participation of Industry and Community Leaders

The establishment of Portland's Energy Policy Steering Committee demonstrates the importance of encouraging consultation and participation of concerned industries and influential community leaders. This broadening of the participative base and increase in perceived acceptability and legitimacy of conservation programs is essential to the process of effective implementation.

Similar results were associated with the 1973 Los Angeles ordinance restricting electricity consumption. Although some of this success can be attributed to patriotic sentiment, the creation of a broad-based committee of business, civic, and labor leaders doubtlessly was a contributing factor.

Ease of Enforcement

It is essential that programs containing mandatory elements also incorporate mechanisms of enforcement that do not lead to excessive bureaucratic complexity or encourage and facilitate consumer avoidance or evasion. An example of a workable enforcement mechanism is Portland's provision for clouded titles in real estate transactions where insulation retrofit has not occurred.

High Coverage and Impact

Programs with a high coverage and impact present the most desirable alternatives to decision makers. In spite of this fact, however, it is surprising how many programs are adopted that violate this simple maxim (Nemetz, 1979). Oregon was particularly prescient in tying weatherization loans to regular homeowner loans for veterans. The Oregon Department of Veterans Affairs processes 25% of all residential mortgages in the state and, in dollar value, is the largest originator of single-family residential mortgages in the United States.

Need for Mandatory Elements

As concluded earlier, the most effective policy of conservation is one that contains a package of complementary measures: economic incentives, information programs, and mandatory elements. Again, experience in Oregon illustrates the validity of this conclusion. Veterans' loans for home purchase are dependent on adequate weatherization, and additional loans are provided for this purpose.

Need for Follow-up Monitoring

One outstanding deficiency associated with several state programs has been an absence of follow-up monitoring and evaluation (Nemetz, 1979). It is often the case that program assessment is impossible because no provision has been made for such evaluation. In contrast, Oregon has an in-place system of follow-up monitoring of alternative energy devices installed with a state tax credit. The monitoring data serve at least two important functions: They provide information on current program success and useful, computerized data that are available for the benefit of future program participants.

Need for Quality Control

Even a well-conceived program for energy conservation can encounter problems of quality control, especially if the program receives widespread and rapid acceptance. Under these circumstances it is necessary to devise techniques that will minimize difficulties associated with faulty construction and installation of weatherization material and alternative energy devices. Oregon and Wisconsin have a requirement for the precertification of solar systems that helps meet the need for quality control. In California, inspectors are employed by several utilities to assess the quality of installed insulation and solar equipment.

9.2.2 Unsuccessful Policy Elements

In addition to the nine positive elements of policy design listed above, it is useful to identify several deficiencies of current conservation programs. These examples represent important lessons to be heeded in the design of future incentive legislation.

Inadequate Monetary Incentives

Programs incorporating low relative or absolute monetary incentives have generally been considered as failures. In Oregon, space-heating customers of publicly owned utilities responded poorly to weatherization programs largely because of the low electricity rates associated with Pacific Northwest hydroelectric power production.

In other states the removal of sales taxes on weatherization material and alternative energy devices has been perceived as providing little incentive to consumers (Nemetz, 1979).

Inadequate Prior Consultation

As successful as Oregon's loan program has become, its initial implementation could have been improved by prior consultation with realty organizations and contractors in the state. Initial resistance to the legislation by these important groups was generated by a concern over the potential delay of commission payments. Apparently, the issue has been largely resolved through the use of belated consultation.

Poor Information Dissemination

Some confusion has arisen in California, Oregon, and several other states over eligibility criteria for weatherization and alternative energy device incentives. In many cases a simpler and broadly disseminated explanation of criteria concerning the timing and nature of installation would have greatly facilitated the conservation process.

Poor Targeting of Program Respondents

The targeting of intended respondents represents one of the classic problems of implementation—a problem that cannot always be anticipated in program design. In Canada and the United States many loan and income tax credit programs designed for low-income people were essentially dysfunctional because the intended respondents did not have sufficient income to capitalize on these incentives (Nemetz, 1979).

In Oregon a weatherization refund of $300 for low-income people did not have a significant impact because the initial process of identification of intended beneficiaries relied on an analysis of state income tax returns, thereby missing a large proportion of the target group.

In California a similar mistargeting occurred with state income tax incentives to encourage the interfuel substitution of solar power for natural gas. An examination of response data indicated that the largest group of program participants were middle- to upper-income people who had changed the method of heating their swimming pools. Despite the program's unintended consequences, however, there are some benefits associated with the results. Insofar as the natural gas would have been consumed for swimming pool heating without the program, a reduction in its use has been achieved.

Wisconsin has successfully tackled the targeting problem with the introduction of its Low Income Weatherization Program, and recent changes in eligibility criteria for its Renewable Energy Program.

9.3   THE NEED FOR BEHAVIORAL DATA

Because of the rapidly changing energy supply and demand situation, it is necessary to gather disaggregated behavioral response data to complement existing econometric studies of price elasticity.

Questionnaires should be developed that can assess the responses of individual consumers to a variety of economic incentives and link these potential responses to important consumer attributes. This methodology promises to increase the predictive power of analyses of consumer response.

Initially, such questionnaires should focus on the two largest areas of consumer energy use: private transportation and home heating. They should also incorporate two types of questions that are sometimes omitted from similar types of research:

1. Questions designed to elicit quantitative (in addition to qualitative) response information; and
2. Questions addressing the issue of perceived program effectiveness as well as acceptability. In other words, consumers are asked not only what incentive mechanism they like, but also what mechanism they think will influence their behavior, regardless of program acceptability.*

*An example of a possible semiquantitative questionnaire is presented as Appendix 2 to this book.

Clearly, there are some problems associated with this type of research as consumers are asked to describe their response to hypothetical situations. This information may be distorted by the respondent's ignorance or by an unwillingness to answer correctly.

Nevertheless, these deficiencies do not detract from the value of such a research program as a complement to econometric data and case studies. These questionnaires should be administered to both consumer panels and random public samples, as each category has particular research value. While the use of a consumer panel affords a unique opportunity to assemble a body of useful and accurate consumer attribute information, the well designed, random public sample guarantees that response data have a high representative and predictive value.

PRICE ELASTICITY OF DEMAND FOR ENERGY

A.1 INTRODUCTION

The "energy crisis" is a relatively recent phenomenon and has resulted in an increasing interest in the analysis of the determinants of demand. Consequently, a significant number of econometric studies on demand at the consumer level have been conducted in the United States during the last decade. A representative bibliography of these studies is included in this appendix. Several Canadian studies have also been conducted, most of them using concepts, models, and methodology developed in the United States but applied in the Canadian context.

The purpose of this appendix is to review some of the existing literature with respect to price elasticities of demand for motor vehicle gasoline, natural gas, electricity, and fuel oil at the residential consumer level and to interpret them in the context of using economic incentives and disincentives to modify energy demand. These energy forms account for the bulk of the energy consumed at this level. Before commencing a review of specific models, important background considerations are presented.

A.2 BACKGROUND

A.2.1 The Derived Demand for Energy

Consumers do not demand energy per se, but the output of capital stocks using energy as an input—for example, heat from furnaces and light from lamps. Hence, energy is a derived demand because demand depends on the existing capital stock, its depreciation rate, additions to it, and its rate of utilization. When dealing with consumer durables, a distinction between stocks already installed and new or replacement stocks must be made in order to separate short-run from long-run energy demand. In the short run, energy demand is tied to "fixed" stocks and is therefore relatively unresponsive to changes in exogenous variables due to the high costs of switching to a device using a different fuel and the difficulty of fuel substitutability. Changes in energy demand in the short run are a result of changes in the utilization rate of existing stocks. In the long run, however, the demand for energy is tantamount to the demand for the energy-using stock itself. Hence, changes in such variables as income or fuel prices over time will lead to a revision of the desired capital stock as well as changes in utilization rates. Econometric models that incorporate the distinction between short-run changes in the utilization rate and the long-run adjustment of capital stocks are called dynamic, whereas those that do not are called static. Long-run elasticities are given directly by dynamic models, but, for static models, elasticities are only considered to be long run on the assumption that a full adjustment in the desired level of capital stocks is made within the period studied. This assumption is a weakness of the static models reviewed.

A.2.2 Energy Supply

All energy demand models reviewed assume that supply is perfectly elastic, which implies that price is an exogenous variable. In many cases (e.g., when rates are regulated), such an assumption is acceptable. In others, however, a more correct, albeit complex, approach would be to include a supply side in the model so that the interaction of supply and demand has an influence in determining the price.

A.2.3 Price Elasticity of Demand

Price elasticity of demand is theoretically defined as the percentage change in the quantity of a good demanded Q divided by a small percentage change in its price P. The general formula is $(\Delta Q/Q)/(\Delta P/P)$, which can be written as $(dQ/dP)(P/Q)$, if the changes are extremely small. Price elasticity is given directly in a multiplicative model and can be derived from additive or exponential models.

For a multiplicative model, suppose energy consumption Q is a function of price P and income Y:

$$Q = \beta_0 P^{\beta_1} Y^{\beta_2} \tag{1}$$

where β_0, β_1, and β_2 are the parameters to be estimated. A multiple linear regression can be performed by converting the function into double logarithmic form:

$$\log Q = b_0 + \beta_1 \log P + \beta_2 \log Y \tag{2}$$

where $b_0 = \log \beta_0$.

To find the price elasticity, one can first take the partial derivative of (1) with respect to P:

$$\frac{\partial Q}{\partial P} = \beta_0 \beta_1 P^{\beta_1 - 1} Y^{\beta_2} \tag{3}$$

Next, multiply both sides of (3) by P/Q.

$$\frac{\partial Q}{\partial P} \cdot \frac{P}{Q} = \frac{\beta_0 \beta_1 P^{\beta_1} Y^{\beta_2}}{Q} = \beta_1 \tag{4}$$

and the price elasticity is the constant β_1, where the last step is based on (1). Hence, for the double logarithmic form, the price elasticity is given directly by the estimated parameter β_1, and it is invariant to the price level.

The term β_1 can be interpreted as follows: When the price increases by a small percentage, income remaining unchanged, then the percentage change in energy consumed is a multiple β_1 of the percentage change in price. This interpretation is not reliable for a relatively large percentage change in the price because such a change introduces an error into the elasticity coefficient as it is defined here.

By contrast, the price elasticity derived from an additive model is not constant but depends on the values of Q and P. Suppose that energy consumption can be modeled using the linear form:

$$Q = \beta_0 + \beta_1 P + \beta_2 Y \tag{5}$$

To obtain the price elasticity, one can take the partial derivative of (5) with respect to P:

$$\frac{\partial Q}{\partial P} = \beta_1 \tag{6}$$

multiply both sides of (6) by P/Q:

$$\frac{\partial Q}{\partial P} \frac{P}{Q} = \frac{P}{Q} \beta_1 \tag{7}$$

where P and Q are the price and consumption means. Hence, price elasticity varies with price and consumption.

The price elasticities derived from an exponential model depend on the price levels. An exponential model of the following form:

$$Q = e^{\beta_0} e^{\beta_1 P} e^{\beta_2 Y} \tag{8}$$

can be transformed into the log-linear model,

$$\log Q = \beta_0 + \beta_1 P + \beta_2 Y \tag{9}$$

To calculate the price elasticity, one can take the partial derivative of (8) with respect to P:

$$\frac{\partial Q}{\partial P} = e^{\beta_0} \beta_1 e^{\beta_1 P} e^{\beta_2 Y} \tag{10}$$

and multiply both sides of (10) by P/Q:

$$\frac{\partial Q}{\partial P} \frac{P}{Q} = \beta_1 \bar{P} \tag{11}$$

where $\bar{P}$ is the mean price. Thus, price elasticities derived from an exponential model depend on the price level.

The price elasticities derived from the general energy consumption function, $Q = f(P,Y)$ can be summarized as follows:

multiplicative model: β_1 (double logarithmic form)
additive model: $\beta_1 \bar{P}/\bar{Q}$ (linear form)
exponential model: $\beta_1 \bar{P}$ (log-linear form)

where $\bar{P}$ and $\bar{Q}$ are the mean energy price and mean energy consumption, respectively, and β_1 is the coefficient for the price variable.

A.2.4   The Price Variable

In practice, energy consumers do not always face a single price but rather a schedule of decreasing marginal prices. Block pricing is commonly used in the electricity and natural gas sectors. For the practical purposes of conducting a statistical analysis, researchers have used either (1) an ex post average price calculated by dividing total revenue accruing from the sale of the energy form by the volume sold or (2) the marginal price taken directly from utility tariff schedules. Taylor (1975, pp. 75–80) makes a strong theoretical argument for incorporating both average and marginal prices in energy demand models. On the one hand, using average price alone introduces simultaneous equations bias because increased consumption results in decreased average price when in fact there is no behavioral justification for this relationship. On the other hand, using marginal price alone is incorrect because, although it is relevant to a consumer's decision to purchase within a block, it does not in itself determine why he consumes in that block as opposed to some other block. After performing a statistical analysis, however, Berndt (1978, p. 23) finds that "least squares estimates with the average intramarginal price variable excluded will typically differ very little from regression estimates with that variable included."

A.2.5   Cross Elasticities of Demand

Consumer behavior theory says that consumers respond to relative price changes by rearranging their bundle of goods in order to maximize utility subject to a budget constraint. It is plausible, therefore, that a relative increase in the price of one form of energy will result in substitution away from that form, especially in the long run. Accordingly, it would be desirable to include in an energy demand model the prices of competing forms of energy to reflect such behavior and to provide measures of price cross elasticities of demand; but not all research analyzing energy elasticities has done so.

A.2.6   Level of Aggregation

For the purpose of analyzing the effectiveness of economic incentives on the energy conservation behavior of individual households, it would be desirable to model energy demand at that level. The vast majority of North American studies, however, aggregate data over much larger sectors—for example, at the state, provincial, and national levels. (One exception is the Wilder and Willenborg study conducted in 1975, which analyzed household electricity demand by using a consumer panel.) Therefore, individual household reaction to changes in income, prices, and other variables is glossed over. The implication is that price elasticity coefficients must be interpreted in an aggregate sense rather than in terms of the response of individual consumers or households.

A.2.7   Reversibility of Price Elasticity Measurements

All energy demand models reviewed are based on historical data. Many include data from the period 1958–1971, when the price of energy to households fell in real terms and the residential use of energy intensified (Berndt, 1977, p. 58). It would appear that there is a causal relationship between energy consumption and price; however, one must appreciate that other unspecified factors could contribute to this inverse correlation. In any case, it would be incorrect to assume that price elasticity measures derived from data representing a period of decreasing real energy prices would necessarily be the same in a time of increasing real energy prices. For example, the large capital investment that consumers and society as a whole have made in the automotive sector in the past suggests that the short-run, and even the long-run, demand functions for gasoline have become more inelastic over time. Hence, the extent of consumer response to a gasoline price increase today would be quite different from an equal decrease in price 10 years ago (Waverman, 1977, p. 6).

Furthermore, real energy prices fell relatively smoothly over the 1958–1971 period; therefore, price elasticity coefficients derived from that period are not representative of significant and abrupt price changes that occur during an energy crisis.

A.2.8   Price Elasticity and Energy Conservation

Consumption of price-inelastic goods is by definition relatively unresponsive to price changes. For example, Dewees, Hyndman, and Waverman (1974) find that per capita gasoline sales in Canada are price inelastic in the short run, with an estimate of -0.08. The implication for the oil companies is that a price increase would be met with a less than proportional decrease in sales and a concomitant increase in total revenue. From the point of view of energy conservation, however, there is an implied potential gasoline saving of 0.08% of the current consumption level in the short run if the price of gasoline is increased by 1%. Based on 1976 Canadian gasoline consumption statistics, this implies a saving of 5.7 million imperial gallons of gasoline, which has a retail value of approximately \$4.8 million using 1978 prices. So, despite being price inelastic, the potential gas saving as a result of a price increase is not insignificant and shows that the labels *price elastic* or *inelastic* are not strictly applicable in the context of energy conservation. The central question is not whether the percentage change in energy consumption is more or less than the percentage change in the energy's price with its resulting impact on revenue, but whether there is a significant induced change in energy consumption at all. The magnitude of the percentage change is important for the purposes of comparing the effectiveness of a price increase for alternative energy forms, but even small percentage responses imply the potential for large absolute savings in any one energy form.

Because of the influence of other variables, the process of trying to predict the magnitude, and even the direction, of a change in the consumption of an energy form in response to a price increase is very complex. Suppose that an economic disincentive is the energy conservation policy under consideration, be it a direct price increase, a tax levy, a fine, or a special rate or charge. Because the consumption of energy depends on many other factors, it is difficult to predict whether the price increase would induce a consumption decrease unless one could forecast exactly how the other variables change. A price elasticity coefficient derived from an econometric model is statistically valid only if, among other things, there is no change in the value of the other variables. In the "real world," however, the explanatory variables are changing simultaneously and in different directions, thereby reducing the chances that a policy will achieve any prespecified target of consumption reduction.

To illustrate some of these complexities, assume that the total demand for a particular energy form is a function of its price, disposable income, population, weather, the price of other energy forms, and the price of energy-using capital stocks. The effect of a price increase for the energy form is confounded by changes in the other variables. For example, an increasing population or a sustained period with extreme hot or cold temperatures induce increases in energy demand. Such variables lie largely or completely outside the control of energy decision makers, and in some cases, the effects of these variables will be significantly greater than the impact of changes induced by governmental policymakers. Uncertainty about the differential impacts of the variables influencing energy demand implies that the policymaker cannot predict with confidence the outcome of using an economic disincentive for reaching his or her objective.

A.2.9 Data Problems

Two major problems are faced by researchers in analyzing energy demand at the residential consumer level. The first is that although it would be preferable to develop an energy demand model for each specific end use (e.g., for lighting and cooking), data availability precludes such fine divisions. It is possible to acquire data on the total electricity used by a household but not on how the electricity is used within the household.

The second major problem is that data for electricity and natural gas are usually on a rate category basis. Frequently, the residential sector includes data on small commercial and industrial users and the commercial category includes data on large residential users and apartment complexes. A similar problem exists with retail gasoline data because service stations do not compile separate data for individual and commercial customers. Under these circumstances, it is difficult to acquire data giving an accurate accounting of residential energy use.

A.2.10 Summary

The major considerations discussed above illustrate that it is unwise to interpret price elasticity estimates without giving consideration to the type of estimation model employed, the other variables which bear upon energy consumption, the level of aggregation and the characteristics of the data. Uncertainty about the accuracy of a specific elasticity measurement caused by the aforementioned complexities can be somewhat mitigated if the results of several alternative models are compared or melded. Such a comparison would reveal a range of elasticities with which one can feel more confident than with any one elasticity measurement. If the circumstances are similar, American and Canadian estimates can be used for confirmatory purposes. Finally, price elasticity estimates are only one of several other inputs, such as consumer surveys and expert opinion, for the economic incentive policy decision. This complementary information reduces the uncertainty associated with sole reliance on elasticity methodology.

A.3 CANADIAN PRICE ELASTICITIES OF DEMAND (WITH SUMMARIES OF RECENT AMERICAN STUDIES)

In this section, price elasticities derived from various econometric models are presented together with a brief review of the data, the methodology, the model, and the qualifications. Finally, in the light of the data and the authors' conclusions, an interpretation of the results is proffered.

A.3.1 Gasoline

D. N. Dewees, R. M. Hyndman, and L. Waverman (1974)

DATA: P.E.I., Nova Scotia, and New Brunswick combined and each of the other seven provinces
annual for 1956–1972 period
pooled cross-section time series

METHOD: ordinary least squares

GENERAL MODEL:

$$G = f(P, Y, URB, PA)$$

where G = sales of gasoline per capita

P = price of gasoline deflated by CPI

Y = real personal disposable income per capita

URB = degree of urbanization

PA = price index of automobiles deflated by CPI

MODEL (1):   static form using current values

$$\log G_t = \quad a + b \log P_t + c \log Y_t + d \log URB_t + e \log PA_t$$

MODEL (2):   static form imposing a four-year distributed lag on P and Y

MODEL (3):   dynamic form using current values

$$\log GF_t = a + b \log P_t + c \log Y_t + d \log URB_t + e \log PA_t$$

where GF = quantity of gasoline sold in current year less 80% of quantity sold in previous year (assumes that because of retirement from the fleet and decreased use with increasing age, current gasoline consumption by those cars remaining from the previous year's fleet will be only 80% of the previous year's consumption. The remaining current demand is attributed to new cars and is referred to as "free demand").

MODEL (4):   dynamic form based on Houthakker, Verleger, and Sheehan study (U.S., 1974)

$$\log G_t = a + b \log P_t + c \log Y_t + d \log G_{t-1}$$

resulting price coefficients are insignificant.

MODEL (5):   static form based on Ramsey, Rasche, and Allen study (U.S., 1975)

$$\log G_t = a + bP_t + c \frac{1}{Yt} + dURB_t + ePA_t$$

Summary results are presented in Table A.1.

QUALIFICATIONS

Model Specification. The researchers have developed a more sophisticated model that incorporates the "fairly complex" technological relationship between vehicle attributes and fuel consumption rates, requires hedonic prices for individual characteristics of the vehicle, and requires a separate demand equation for each characteristic (p. 16). A simpler analysis was conducted by the authors because of time and data constraints. If the sophisticated model is correctly specified, then the simpler model must to some extent be misspecified (p. 28). For example, the price of other transportation modes like transit systems is an important variable and is probably inadequately reflected by the degree of urbanization variable.

Gasoline Sales Data. The gasoline sales data include sales for commercial purposes, including farming. Demand for these purposes depends on factors different from the automotive demand factors and there-

TABLE A.1.   Price Elasticities of Demand for Gasoline Sales per Capita 1956–1972[a]

	Model (1) Long Run	Model (2) Long Run	Model (3) Short Run[c]	Model (3) Long Run	Model (5) 1972 only Long Run
Newfoundland	−0.97	−0.99	−0.12	−0.60	−0.97
Maritimes	−0.84	−0.84	−0.10	−0.50	−0.59
				(−1.9)	
Quebec	−0.84	−0.82	−0.10	−0.50	−0.69
Ontario	−0.79	−0.78	−0.10	−0.50	−0.54
Manitoba	−0.80	−0.79	−0.10	−0.48	−0.52
				(−1.9)	
Saskatchewan	−0.74	−0.76	−0.09	−0.44	−0.15
				(−1.7)	(−0.14)
Alberta	−0.74	−0.74	−0.09	−0.44	−0.25
				(−1.7)	
British Columbia	−0.82	−0.82	−0.11	−0.53	−0.58
Canada[b]	−0.78	−0.73	−0.08	−0.40	−0.69
				(−1.5)	

[a]Any t-ratio that is less than 2 is shown in parentheses. All other t-ratios are greater than 2.
[b]For each model separate regressions were performed for Canada allowing individual constant terms for each province, as opposed to allowing individual price coefficients for each province.
[c]Given that the "free demand" is assumed to be 20% of the total demand, the short-run price elasticity is one-fifth of corresponding long-run value.

fore their inclusion will introduce some error (p. 28). This inclusion might explain the prevalence of insignificant t-ratios for the prairie provinces where a significant amount of gasoline is sold for farm use.

Price Data.   The price data used were obtained from a major oil company and do not reflect the discount prices that were offered by some retail outlets. Hence, the price elasticities are probably understated.

INTERPRETATION

Ignoring the complexities already discussed in Section A.2, a 1% increase in the price of gasoline in Canada would have the potential of decreasing per capita gasoline sales by about 0.08% and 0.70% in the short and long run, respectively. Applying these price elasticities to the 310 gallons per capita Canadian gasoline consumption in 1976, the hypothetical price increase of 1% implies the potential for a consumption decrease of 0.25 gallons per capita in the short run, or a total of about 5.748 million imperial gallons. Using average 1976 retail prices of regular gasoline for specified cities in Canada, this volume is equivalent to a consumer expenditure of about $4.8 million. As mentioned in Section A.2, however, some caution must be used in interpreting this implied decrease in consumption.

Ake G. Blomqvist and W. Haessel (1978)

DATA: eight Canadian provinces (i.e., excluding Manitoba and Saskatchewan)
1971–1975 period
pooled cross-section time series

METHOD: ordinary least squares (OLS)
two-stage least squares (TSLS)
variance components (VC)

GENERAL MODEL:

$$S^*_{ai} = F(P, \Pi_m, Y, D)$$

where S^*_{ai} = aggregate desired demand for the stock of cars in age class a and size class i

P = a vector of car prices

Π_m = index of car mileage costs represented by the price of gasoline based on regional city indices adjusted for regional differences

Y = per capita disposable income

D = vector of demographic characteristics

MODEL VARIATIONS: dynamic models are developed that allow a focus on either the determinants of automobile stock or automobile sales

Summary results are presented in Table A.2.

TABLE A.2. Short-Run Elasticities of Automobile Demand with Respect to the Price of Gasoline (OLS) (Blomqvist and Haessel, 1978, p. 485)

	Elasticity	t-ratio ($n = 40$)
New large car stocks[a]	−0.29	2.65
New small car stocks	−0.0078	0.07
New large car sales	−0.63	2.49
New small car sales	+0.18	0.78

[a]New stocks are cars up to and including three years old.

QUALIFICATIONS

Dependent Variable. The dependent variable of the model is car sales by age and size category, not gasoline demand. Thus the elasticities presented are not the own-price elasticities of cars or gasoline, but the cross-price elasticities of cars with respect to gasoline.

Data. Because of the nature of their data and estimation technique, the authors do not believe that they have estimated accurately the parameters representing the speed of adjustment in the stocks of cars to changing circumstances (p. 488).

INTERPRETATION

The cross-price elasticities of new small car stock and sales with respect to the price of gasoline are based on regression coefficients which are not significantly different from zero (at the 0.01 level) (p. 485). Thus, it is implied that one can be confident that a gasoline price increase would not cause a decrease in the stock or sales of new small cars. On the other hand, a gasoline price increase is predicted to have an impact on the stock and sales of new large cars. Although the values -0.29 and -0.63 imply inelastic demand for new large car stocks and sales, respectively, it is clear from the data that new large car sales are more responsive in the short run to gasoline price increases than new large car stocks. The sales response is the expected and desired result in the case where policymakers wish consumers to purchase small rather than large cars to encourage energy conservation. Unfortunately, in the case of stocks, if gasoline price increases are relatively small, one must be prepared to wait for already purchased cars to run their normal life span before a switch to small cars will occur.

Summary of Recent American Studies

Author	Model Type	Gasoline Price Variable	Estimated Price Elasticity of Demand for Gasoline	
			Short Run	Long Run
1. Lady (1974)	Static single equation; national; monthly 1969–1973	Retail price, excl. taxes	-0.16	N/A
2. Verleger (1973)	Dynamic single equation; state; quarterly 1963–1972	Retail price, excl. taxes	-0.16	-0.54

Author	Model Type	Gasoline Price Variable	Estimated Price Elasticity of Demand for Gasoline	
			Short Run	Long Run
3. Houthakker and Kennedy (1973)	Dynamic single equation; state; annual	Retail price incl. taxes	−0.07	−0.24
4. Anderson (1972)	Dynamic single equation; state; annual 1952–1972	Retail price incl. taxes	−0.11	−0.60
5. Chamberlain (1973)	Dynamic single equation; national; annual 1951–1970	Retail price excl. taxes	−0.06	−0.07
6. McGillivray (1974)	Dynamic single equation; national; annual 1951–1969	Retail price incl. taxes	−0.22	−0.69
7. Houthakker, Verlager (1973)	Dynamic single equation; state; annual 1951–1971	Retail price incl. taxes	−0.43	−0.75
8. Ramsey, Rasche, Allen (1975)	Static demand-supply; national; annual 1948–1971	Retail price incl. taxes	N/A	−0.77
9. Burright et al. (1974)	Static recursive; national; annual 1954–1972	CPI for gas and oil	−0.14	−0.93
10. Sweeney (1975)	Dynamic recursive; national; annual 1951–1973	Retail price incl. taxes	−0.12	−0.73

Source: Kraft and Rodekohr, 1978, p. 53.
Notes: N/A, not available.

A.3.2 Natural Gas

G. C. Watkins (1974)

DATA: British Columbia, Ontario, and Quebec
annual for 1959–1972 period
time series pooled with estimated coefficients for
the intercept, the income variable, and the lagged
gas consumption variable

TABLE A.3. Price Elasticity of Demand for Total Residential and Commercial Natural Gas Consumption

| | Total Demand (captive + flexible)[a] | | Flexible Demand, Linear Model |
	Linear Model	Double-Log Model	
Quebec	−0.45	−0.28	−2.47
Ontario	−0.24	−0.30	−1.36
British Columbia	−0.32	−0.29	−1.66

Source: Watkins, 1974, p. 23.

[a]Captive plus flexible demand is equivalent to the short-run demand, whereas flexible demand is equivalent to the long-run demand.

METHOD: ordinary least squares

GENERAL MODEL: dynamic

$$G_t = a_0 + a_1 P_t + a_2 H_t^* + a_3 Y_t^* + a_4 G_{t-1}$$

where G_t = total residential and commercial gas consumption in Btu (10^{12}), year t

P_t = index of ratio of price of gas to price of fuel oil, year t (1967 = 100)

$H_t^* = H_t - (1 - r)H_{t-1}$

$Y_t^* = Y_t - (1 - r)Y_{t-1}$

H_t = number of households, year t (thousands)

Y_t = real personal disposable income per household, year t (1961 \$)

r = depreciation rate for total fuel market

Summary results are presented in Table A.3.

QUALIFICATIONS

Dependent Variable. Gas consumption includes commercial; therefore the results are not strictly applicable to the residential sector.

Price Variable. Because the price variable is in ratio form, the elasticities presented are neither own-price nor cross-price elasticities. The ratio form avoids the problem of having to deflate the price series to define real prices, it reduces collinearity if the two-price series are correlated, and it "eliminates the question of an efficiency adjustment for the oil price vis-à-vis the gas price as long as the adjustment does not vary appreciably over time" (p. 13). Natural gas prices are based on value-volume data and hence are average rather than marginal prices.

Depreciation Rate.   The model's depreciation rate is fixed, whereas it is known that the rate varies with the changing age structure of equipment and depends on the effect of changing circumstances on obsolescence (pp. 14–15). Given these complexities and a lack of data, no attempt was made to estimate depreciation rates.

Provinces Used.   Formulation of the model assumes a reasonable degree of interfuel competition. In the Prairies, once gas service is available, gas is predominant. In this area, gas prices have been consistently less than fuel oil prices and competition is minimal. Accordingly, the model was applied solely to British Columbia, Ontario, and Quebec, where competition is more prevalent (p. 18).

Quebec Elasticities.   The absolute values of the Quebec natural gas price elasticities are almost double the Ontario values when using the linear model. This difference occurs only when using a particular functional form, and for this reason the result must be regarded with suspicion. Other factors that might explain the difference are that market penetration has been discouraged in Quebec because of excessive combustion efficiency (i.e., explosions), a tendency for gas premiums to be wider than in other provinces, and the fact that some of the market consists of single heaters rather than central heating equipment (p. 24).

INTERPRETATION

As expected, flexible or long-run demand is more price elastic than total or short-run demand, implying that energy conservation is more difficult to encourage in the short run than in the long run, at least when using price increases as the mechanism. A 1% increase in the price of natural gas would have the potential of decreasing total residential and commercial gas consumption in Ontario and British Columbia by about 0.32% and 1.83% in the short and long runs, respectively. These estimates can also be applied to Quebec, although, as discussed in the qualifications, with less certainty.

E. R. Berndt and G. C. Watkins (1977)

DATA:	British Columbia and Ontario
	annual for 1959–1974 period
METHOD:	Model A: maximum likelihood
	Model B: ordinary least squares

GENERAL MODEL: dynamic

$$G_t = DD_t^{\alpha_6}[\alpha_0 + P_t^{\alpha_1}H_t^{*\alpha_2}Y_t^{*\alpha_3}] + \left[\frac{DD_t}{DD_{t-1}}\right]^{\alpha_6} \alpha_5\, G_{t-1}$$

where G_t = total residential and commercial gas consumption in period
        t.

  DD_t = ratio of degree days (i.e., the product of the number of de-
        grees below 65°F or 18°C and the number of days) in year t
        to average degree-days.

  P_t = index of the ratio of weighted residential and commercial
        gas price to weighted residential and commercial fuel oil
        price (1967 = 100).

  H_t^* = number of new households (thousand) in year t

  Y_t^* = new real disposable income per household (thousands of
        1961 \$) in year t

  α_6 = can be constrained to equal 0 or 1 so that one is able to test
        the hypotheses that degree-days is an insignificant variable
        or that a conventional degree-day adjustment factor is ap-
        propriate (As it turns out, degree-days is a significant vari-
        able).

Summary results are presented in Table A.4.

QUALIFICATIONS

  Dependent Variable.   Gas consumption includes commercial; therefore the results are not strictly applicable to the residential sector.

**TABLE A.4.   Price Elasticities of Demand for Total
Residential and Commercial Gas Consumption**

	Model A		
Short Run	B.C.	Ontario	Model B
1959	−0.25	−0.25	not estimated
1962	−0.18	−0.16	not estimated
1965	−0.19	−0.13	not estimated
1968	−0.14	−0.13	not estimated
1971	−0.16	−0.12	not estimated
1974	−0.15	−0.12	not estimated
Long Run	−0.686		−0.733

Source: Berndt and Watkins, 1977, pp. 104 and 106.

Price Variable.   Because the price variable is in ratio form, the elasticities presented are neither own-price nor cross-price elasticities. The ratio form avoids the problem of having to deflate the price series to define real prices, it reduces collinearity if the two-price series are correlated and "eliminates the question of an efficiency adjustment for the oil price vis-à-vis the gas price as long as the adjustment does not vary appreciably over time" (Watkins, 1974, p. 13). Natural gas prices are based on value-volume data and hence are average rather than marginal prices.

Depreciation Rate.   The model's depreciation rate is fixed, whereas it is known that the rate varies with the changing age structure of equipment and depends on the effect of changing circumstances on obsolescence. Given these complexities and a lack of data, no attempt was made to estimate depreciation rates.

INTERPRETATION

Just as Watkins had shown in his 1974 study, "variations in price have little impact on gas demand in the short run but have a more substantial impact over the long run" (Berndt and Watkins, 1977, p. 108). The price elasticity values in the Berndt-Watkins model, however, are more inelastic for both the short and long run than the values derived from the 1974 Watkins model. The Berndt-Watkins model produces price elasticities for Ontario and British Columbia of about -0.15 and -0.70 in the short and long run, respectively, versus the estimates of -0.32 and -1.83 that Watkins had found earlier. One explanation for this difference might be the introduction of degree-days as an independent variable. Its statistical significance is plausible given that the predominant portion of residential and commercial demand is for space heating. As a result of the low price elasticity of demand, gas conservation in this sector is relatively difficult, especially in the short run, because of the impact of inclement weather on natural gas consumption in Canada.

Berndt and Watkins give three reasons for the short-run price elasticities' decline (in absolute value) over time. First, the decline is partly due to the mathematical structure of the model (p. 105, fn. 19). Second, "time may be a surrogate for factors such as rising real income, which may reduce cost consciousness and hence response intensities" (p. 105). And third, the decline "is consistent with the concept of rapid threshold development in the early stages of the introduction of gas supplies, followed by a maturing in the rate of penetration as the initial impetus recedes" (p. 106). Given that real incomes continue to rise in British Columbia and Ontario, one can extrapolate that the second and third reasons imply continued downward pressure on elasticities. One can conclude from this phenomenon that, if the use of economic incentives or disincentives to encourage energy conservation is postponed, they will be less effective. Increasing real incomes and huge investments in energy infrastructure like pipelines or even household energy-using equipment make economic incentives or disincentives less attractive and more difficult to respond to, especially in the short run, unless they are sufficiently large to override these constraints.

M. Fuss and L. Waverman (1975); and M. Fuss, R. Hyndman, and L. Waverman (1977)

DATA: regional for Atlantic provinces, Quebec, Ontario, the Prairies, and B.C.

1958–1971 period

pooled cross-section time series

METHOD: ordinary least squares

GENERAL MODELS: static

MODEL A: double-logarithmic form (1975)

$$\log GPH = a + b \log PGO + c \log YH + d \log YHQ$$
$$+ e \log SM + f \log DDRN + g \log DQ$$
$$+ h \log DO + i \log DP + j \log DB$$

MODEL B: exponential form with inverse of income variables (1975)

$$\log GPH = a + b\,PGO + c\,IYH + d\,IYHQ + e\,SM$$
$$+ f\,DDRN + g\,DQ + h\,DO + i\,DP + j\,DB$$

(1977 version is the same except that $IYHQ$ is excluded)

where GPH = aggregate of natural gas and petroleum products consumption per household

PGO = aggregate price of gas and oil deflated by the Consumer Price Index using estimated shares as weights for aggregation

YH = personal disposable income per household

YHQ = YH slope dummy for Quebec

IYH = inverse of personal disposable income per household

$IYHQ$ = IYH slope dummy for Quebec

SM = single-family dwellings as a proportion of total dwellings

$DDRN$ = degree days below 65°F relative to normal

DQ, DO, DP, and DB = intercept dummies for Quebec, Ontario, the Prairies, and B.C., respectively

Summary results are presented in Table A.5.

QUALIFICATIONS

Level of Aggregation. Both the dependent variable and the price variable aggregate over natural gas and petroleum products. Hence, the

TABLE A.5. Long-Run Price Elasticities of Demand for the Aggregate of Natural Gas and Petroleum Products Consumption

Regions	Model A		Model B	
	Per Household	Per Capita	Per Household	Per Capita
Atlantic	−0.86	−0.96	−0.89	−0.93
Quebec	−0.86	−0.96	−0.81	−0.85
Ontario	−0.86	−0.96	−0.82	−0.86
Prairies	−0.86	−0.96	−0.53	−0.56
B.C.	−0.86	−0.96	−0.71	−0.74
Canada (average of regional values)	−0.86	−0.96	−0.75 (−0.733[a])	−0.79

Source: Fuss and Waverman, 1975, p. 118, 120.
[a]Fuss, Hyndman, and Waverman, 1977, p. 164.

price elasticities are not strictly applicable to either one or the other alone.

Static Model. As a static model, the dynamic nature of energy demand and equipment depreciation is not accounted for. Full adjustments in demand are assumed to take place within a year and the elasticities are accordingly interpreted as long run.

Interfuel Substitution. The model assumes no substitution between electricity and the fossil fuels, which is appropriate for some household uses of energy but not for others. Accordingly, the model is misspecified for the aggregate of household uses (1975, p. 90).

Conversion Efficiency. "The difference in conversion efficiencies for different types (of fuels) was allowed for by converting consumption and prices to output BTU terms" (1975, p. 94). Fuel consumption was multiplied by a conversion efficiency ratio while fuel price was divided by the same ratio.

Price Variable. The natural gas price is the marginal price at rates of consumption for a typical customer using gas for heating.

INTERPRETATION

Fuss, Hyndman, and Waverman state in their 1977 study that "for fossil fuels, an inelastic price response on the part of residential users is found" (1977, p. 167). In terms of energy conservation, however, the model implies a significant consumption decrease in response to price increases. The long-run consumption of fossil fuels per household is predicted to decrease by about 0.8% in response to a 1% increase in the fuels' prices. An important qualification, however, is that the model is static and, as such, merits less confidence than dynamic models.

G. F. Mathewson and Associates for Ontario Hydro (1976)

DATA: regional for Quebec, Ontario, the Prairies, and B.C.
                  1958–1971 period
                  pooled cross-section time series
METHOD: ordinary least squares
GENERAL MODEL:

$$\Delta E^g = (a + b\,P_e + c\,P_o + d\,P_g)\epsilon R + eH(-1)R\,(-1) + f\,\Delta D + H\,\Delta g_a R$$

where ΔE^g = a one-year change in total residential natural gas consumption

P_e = the marginal price of electricity for residential consumers between 500 kWh and 1000 kWh per month (deflated by the consumer price index, CPI), lagged one year

P_o = the retail price of home heating oil (deflated by the CPI), lagged one year

P_g = the marginal price of gas to a typical residential heating customer (deflated by the CPI), lagged one year

R = proportion of households with access to natural gas

H = number of households

D = degree-days below 65°F in the months of September to May

g_a = "typical" natural gas appliance load by a household which is defined to be a function of fuel prices, durable good costs, and income

Summary results are presented in Table A.6.

TABLE A.6.   Price Elasticities of Demand for Total Residential Natural Gas Consumption

	Heating		Appliance
	Upper Bound	Lower Bound	(Long Run)
P_e (marginal price of electricity)	3.70	2.53	0.51
P_o (retail price of home heating oil)	2.68[a]	2.05[a]	NE
P_g (marginal price of gas)	−1.77	−1.35	−0.21

NE (not estimated)
Source: G. F. Mathewson and Associates, 1976, p. 52.
[a]*t*-ratio is insignificant.

QUALIFICATIONS

Separation of Household Decisions. The model assumes that heating decisions are separate from and independent of appliance decisions. The separation is assumed because the former requires "lumpy" investments and the latter are decisions of a more continuous nature. There is no guarantee that the model's assumption holds.

Durable Goods and Energy Use. In the production of services from energy and durable goods by households, it is assumed that technological coefficients are fixed (p. 31).

This assumption permits the aggregation of units of durables and units of energy into an "energy package," g_a. The coefficient is the technologically optimal mcf* per unit of gas appliance. This assumption, which does not allow for variation in the intensity of appliance use, might be unrealistic, however, because it can be anticipated that energy price changes and income changes have an effect on the rate of appliance usage.

INTERPRETATION

In the words of the authors, the "empirical estimates indicate that major changes in the heating system of houses are very price and income responsive, whereas appliance decisions of a more continuous nature are much less price and income responsive with stocks adjusting fairly quickly towards their long-run levels" (G. F. Mathewson and Associates, 1976, p. 43). These estimates cannot be compared directly to the other natural gas price elasticity estimates reviewed because of the separation of the heating and appliance decisions in the model. The relatively inelastic own-price estimate for appliances, however, is consistent with the intuitive conclusion that a significant energy price increase would be necessary before a consumer would decide to switch to appliances using a more inexpensive fuel, especially since the proportion of total energy used by a household's appliances is much less than that for the household's heating system.

The relatively elastic own-price estimates for heating are also plausible because in the construction of new homes or in major renovations, some thought would be given to the type of heating fuels available and their relative prices. It can be inferred from these estimates that a policymaker could use the price mechanism to influence the type of fuel adopted for heating purposes.

*thousand cubic feet

Robert S. Pindyck (1976)

DATA: Belgium, Canada, France, Italy, the Netherlands, Norway, United Kingdom, United States, and West Germany

pre-1974

international pooled cross-section time series

METHOD: iterative nonlinear Zellner estimation with country dummy variables

GENERAL MODEL: static version of indirect translog utility function

$$S = f(P_i/M, t)$$

where S = budget shares expended on natural gas

P_i/M = ratio of the countrywide average of the average retail price of fuel i (measured in local currency units per teal and converted to 1970 U.S. dollars) to total residential consumption expenditure on energy

t = time

Summary results are presented in Table A.7.

QUALIFICATIONS

Static Model. As a static model, lagged responses to a sudden price change are not incorporated, but rather are assumed to occur instantaneously (p. 14). Therefore, Pindyck's elasticities are assumed to be long run.

Homotheticity. A homotheticity assumption is imposed on the model because Pindyck had difficulty in computing nonhomothetic

TABLE A.7. Long-Run Price Elasticities of Demand for Natural Gas (Pindyck's Model 7)

Own-price	1962	−1.17
	1968	−1.15
Cross-price		
Liquid fuel	1962	0.19
	1968	0.17
Electricity	1962	−0.29
	1968	−0.26

Source: Pindyck, 1976, p. 62

models (p. 55). Under homotheticity, the budget share S is independent of total expenditures M (p. 9). According to Pindyck, the qualification is critical enough to render the results inconclusive.

Capital Stock.   The model assumes that the expenditure share on a fuel is independent of the expenditure shares on the energy-using capital stocks.

Data and Results.   The quality of the data varies among countries and the results in general are preliminary and incomplete.

Price Variable.   The price of natural gas is an ex post average price.

Cross-Price Elasticity.   The negative cross-price elasticity between natural gas and electricity appears contradictory to the theory. Substitution effects between the fuels were probably overridden by, as Pindyck says, the simultaneous installation of new gas lines and electric power lines during the 1960s in many European countries, together with a diminishing use of coal (p. 60). It can therefore be inferred that consumption of each fuel increased because of its increased availability irrespective of real price decreases in the other fuel.

INTERPRETATION

Quality and age of the data, the model's static formulation and the homotheticity assumption are a few of the reasons why the price elasticities are suspect. Despite these caveats, the own-price elasticities are still "in the ball park" with the other estimates reviewed for natural gas.

The relatively small cross-price elasticities (in absolute value) with respect to liquid fuel and electricity suggest that the degree of substitution between natural gas and each of the other energy forms is small. The credibility of these cross-price elasticities as representative of the long run, however, is questionable because the static formulation of the model does not recognize lagged demand responses to a price change as equipment becomes obsolete and is replaced. Should we accept that the highly suspect cross-price estimates are indeed representative of the "true" values and disregard several other qualifications, then it is implied that, in the long run, price increases in liquid fuels or electricity would not be totally manifested in consumption increases in natural gas but rather in the continued use of each of the other fuels despite price increases, or in an absolute decrease in the amount of total energy consumed. (The small cross-price elasticities between electricity and liquid fuel imply that switching between them is also unlikely.)

Although it is questionable that the cross-price elasticities are representative of the long run, we can accept them as being "an upper bound" for the short-run value with some confidence, especially since the model is a static formulation. Small short-run elasticities can be explained by the fact that consumers are tied to energy-using equipment.

The implication is that price increases in electricity or fuel oil implemented to encourage conservation of those fuels probably will not result in a significant consumption increase in natural gas, especially in the short run.

Summary of Recent American Studies

Author	Type of Data	Type of Price	Price Elasticity of Demand for Natural Gas	
			Short Run	Long Run
Houthakker and Taylor (1970)[a]	TS: aggregate U.S.	A	0	0
Anderson (1973)[a]	CS: states	A[+]	NE	−1.73
Erickson, Spann, and Ciliano (1973)[b]	CS	A	NA	−1.47
Randall, Ives, and Ryan (1974)[a]	CS: communities in Southwest	A	NE	−1.12
Joskow and Baughman (1975)[b]	CS: TS	A	NA	−0.62[c]
FEA (1976d)[a]	CS: TS census regions	A	−0.16	−1.26

Notes: CS, cross-section; TS, time series; A, ex post average price; A[+], price for a fixed number of therms; NE, not estimated; NA, not applicable.

[a]Cited in Taylor, 1977, p. 21.

[b]Cited in Waverman, 1977, p. 7.

[c]Nonlinear equation, elasticity calculated at mean.

A.3.3   Electricity

M. Fuss and L. Waverman (1975); and M. Fuss, R. Hyndman, and L. Waverman (1977)

DATA: regional for Atlantic provinces, Quebec, Ontario, the Prairies, and B.C.

1958–1971 period

pooled cross-section time series

METHOD: ordinary least squares

GENERAL MODELS: static

MODEL A: double logarithmic form (1975)

$$\log EH = a + b \log APE + c \log YH + d \log YHQ$$
$$+ e \log SM + f \log DQ + g \log DO$$
$$+ h \log DP + i \log DB$$

MODEL B: exponential form with inverse of income variables

1975: $\log EH = a + b\,APE + c\,IYH + d\,IYHQ + e\,SM +$
$$f\,DQ + g\,DO + h\,DP + i\,DB$$

1977: $\log EH = a + b\,MPE + c\,IYH + d\,IYHQ + e\,SM$
$$+ f\,DDRN + g\,DQ + h\,DO + i\,DP + j\,DB$$

where EH = electricity consumption per household

APE = average price of electricity

MPE = marginal price of electricity

YH = personal disposable income per household

YHQ = YH slope dummy for Quebec

IYH = inverse of personal disposable income per household

$IYHQ$ = IYH slope dummy for Quebec

$DDRN$ = heating degree days relative to normal

SM = single-family dwellings as a proportion of total dwellings

DQ, DO, DP, and DB = intercept dummies for Quebec, Ontario, the Prairies, and B.C., respectively

Summary results are presented in Table A.8.

QUALIFICATIONS

Static Model.   As a static model, the dynamic nature of energy demand and equipment depreciation is not accounted for. Full adjustments

TABLE A.8.   Long-Run Price Elasticities of Demand for Electricity Consumption

Regions	Model A		Model B	
	Per Household	Per Capita	Per Household	Per Capita
Atlantic	−0.29	−0.34	−0.22	−0.24
Quebec	−0.29	−0.34	−0.18	−0.20
Ontario	−0.29	−0.34	−0.16	−0.17
Prairies	−0.29	−0.34	−0.15	−0.17
B.C.	−0.29	−0.34	−0.24	−0.27
Canada (average of regional values)	−0.29	−0.34	−0.19 (−0.14[a])	−0.21

Source: Fuss and Waverman, 1975, p. 118 and 120.
[a]Fuss, Hyndman, and Waverman, 1977, p. 164.

in demand are assumed to take place within a year and the elasticities are accordingly interpreted as long run.

Interfuel Substitution. The model assumes no substitution between electricity and fossil fuels, which is appropriate for some household uses of energy but not for others. Accordingly, the model is misspecified for the aggregate of household uses (1975, p. 90).

Conversion Efficiency. "The difference in conversion efficiencies for different types (of fuels) was allowed for by converting consumption and prices to output BTU terms" (1975, p. 94). Fuel consumption was multiplied by a conversion efficiency ratio while fuel price was divided by the same ratio.

Price Variable. The electricity price variable is different for the 1975 and 1977 models. In 1975 an average price was used, whereas a marginal price was used in 1977.

INTERPRETATION

The authors accept the low price elasticities as plausible because electricity is used mainly for lighting, cooking and the like; "uses not rapidly reduced as price increases" (Fuss, Hyndman, and Waverman, 1977, p. 167). Of the four energy forms reviewed in this report, electricity is the most price inelastic in the long run and, by deduction, among the most inelastic in the short run. There appears to be little "slack" in the amount of electricity used, implying that there is little potential for significant per capita consumption curtailment of this fuel. This inference is especially credible when electricity consumption is compared to gasoline consumption. Dewees, Hyndman, and Waverman (1974) found that the short-run demand for gasoline was also quite inelastic at -0.08. The sheer magnitude of motor vehicle gasoline consumption as compared to residential electricity consumption, however, implies that a percentage decrease in per capita gasoline consumption as a result of some energy conservation measure would be much more significant than a similar percentage decrease in per capita electricity consumption. In 1976, motor vehicle gasoline consumption in Canada was about five times greater than residential electricity consumption in terms of Btu's (Statistics Canada, 1978a and 1978b). Furthermore, it is clear that there is relatively more "slack" in gasoline consumption. Use of alternative modes of transportation like transit systems and curtailment in the amount of "pleasure" driving would produce significant energy savings without imposing undue hardships on the consumers' "quality of life" and family budget. As for electricity, however, consumer "quality of life" and financial costs would be higher for an equivalent per capita Btu consumption decrease. Unless more energy efficient cooking equipment and lighting fixtures are developed and purchased, consumers would have to decrease their use of cooking equipment and lighting.

G. F. Mathewson and Associates for Ontario Hydro (1976)

DATA: regional for Quebec, Ontario, the Prairies, and B.C.
                     1958–1971 period
                     pooled cross-section time series
METHOD: Model A: ordinary least squares
                     Model B: Hildreth-Lu iterative
GENERAL MODEL:

$$\Delta E^e = (a + b\,P_e + c\,P_o + d\,P_g)C + eH(-1) + f\Delta D + H\Delta e_a$$

where ΔE^e = a one-year change in total residential electricity consumption

P_e = marginal price of electricity for residential consumers between 500 kWh and 1000 kWh per month (deflated by the consumer price index, CPI) lagged one year

P_o = the retail price of home heating oil (deflated by the CPI) lagged one year

P_g = the marginal price of gas to a typical residential heating customer (deflated by the CPI) lagged one year

C = housing completions measured as a two-year moving average of single-family, semidetached, and row housing completions weighted by an index of size for heating requirements

H = number of households

D = degree days below 65° F in the months of September to May

e_a = "typical" electric appliance load by a household which is defined to be a function of fuel prices, durable good costs, and income

Summary results are presented in Table A.9.

QUALIFICATIONS

Separation of Household Decisions.   The model assumes that heating decisions are separate from and independent of appliance decisions. The separation is assumed because the former requires "lumpy" investments and the latter are decisions of a more continuous nature. There is no guarantee that the model's assumption holds.

TABLE A.9. Price Elasticities of Demand for Total Residential Electricity Consumption

| | Heating | | Appliance |
	Upper Bound	Lower Bound	(Long Run)
Model A			
P_e (marginal price of electricity)	-6.07	-2.28	-0.01
P_o (retail price of home heating oil)	-5.00^a	-1.88^a	NE
P_g (marginal price of gas)	3.73	1.40	-0.07^a
Model B			
P_e	-6.86	-1.20	-0.10
P_o	-2.29^a	-0.40^a	NE[b]
P_g	4.71	0.83	-0.07^a

Source: G. F. Mathewson and Associates, 1976, p. 51.
[a]t-ratio is insignificant.
[b]NE, not estimated.

Durable Goods and Energy Use. In the production of services from energy and durable goods by households, it is assumed that technological coefficients are fixed (p. 31). This assumption permits the aggregation of units of durables and units of energy into an "energy package," e_a. The coefficient is the technologically optimal kWh per unit of electric appliance. This assumption, which does not allow for variation in the intensity of appliance use, might be unrealistic, however, because it can be anticipated that energy price changes and income changes have an effect on the rate of appliance usage.

INTERPRETATION

As in G. F. Mathewson and Associates' natural gas model, the "empirical estimates indicate that major changes in the heating system of houses are very price and income responsive, whereas appliance decisions' of a more continuous nature are much less price and income responsive with stocks adjusting fairly quickly towards their long-run levels" (G. F. Mathewson and Associates, 1976, p. 43). The relatively low, own-price elasticities for appliances are consistent with Fuss, Hyndman, and Waverman's (1977, p. 167) estimates, which were explained by the observation that electricity is used mainly for lighting, cooking, and the like; "uses not rapidly reduced as price increases."

Robert S. Pindyck (1976)

DATA: Belgium, Canada, France, Italy, the Netherlands, Norway, United Kingdom, United States, and West Germany pre-1974

international pooled cross-section time series

METHOD: iterative nonlinear Zellner estimation with country dummy variables

GENERAL MODEL: static version of indirect translog utility function

$$S = f(P_i/M, t)$$

where S = budget shares expended on electricity

P_i/M = ratio of the countrywide average of the average retail price of fuel i (measured in local currency units per teal and converted to 1970 U.S. dollars) to total residential consumption expenditure on energy

t = time

Summary results are presented in Table A.10.

QUALIFICATIONS

Static Model. As a static model, lagged responses to a sudden price change are not incorporated, but rather are assumed to occur instantaneously (p. 14). Therefore, Pindyck's elasticities are assumed to be long run.

Homotheticity. A homotheticity assumption is imposed on the model because Pindyck had difficulty in computing nonhomothetic models (p. 55). Under homotheticity the budget share S is independent of total expenditures M (p. 9). According to Pindyck, the qualification is critical enough to render the results inconclusive.

Capital Stock. The model assumes that the expenditure share on a fuel is independent of the expenditure shares on the energy-using capital stocks.

TABLE A.10. Long-Run Price Elasticities of Demand for Electricity (Pindyck's Model 7)

Own-price	1962	−0.80
	1968	−0.81
Cross-price		
Natural gas	1962	−0.10
	1968	−0.10
Liquid fuel	1962	0.08
	1968	0.07

Source: Pindyck, 1976, p. 62.

Data and Results.   The quality of the data varies among countries, and the results in general are preliminary and incomplete.

Price Variable.   The price of natural gas is an ex post average price.

Cross-Price Elasticity.   The negative cross-price elasticity between natural gas and electricity appears contradictory to the theory. Substitution effects between the fuels were probably overridden by, as Pindyck says, the simultaneous installation of new gas lines and electric power lines during the 1960s in many European countries, together with a diminishing use of coal (p. 60). It can therefore be inferred that consumption of each fuel increased because of its increased availability irrespective of real price decreases in the other fuel.

INTERPRETATION

Quality and age of the data, the model's static formulation and the homotheticity assumption are a few of the reasons why the price elasticity estimates are suspect. This discussion of Pindyck's natural gas elasticity estimates apply to the electricity model estimates as well.

Summary of Recent American Studies

Author	Type of Data	Type of Price	Price Elasticity of Demand for Natural Gas	
			Short Run	Long Run
Acton et al. (1976)	CS: small geographical areas	M	(−0.70)	
Acton et al. (1976)	CS:TS: small geographical areas	M	(−0.34)	
Taylor et al. (1976)	CS:TS: states, annual	M	−0.07	−0.81
Lacy and Street (1975)	TS: area served by one utility	M	(−0.45)	
Wilder and Willenborg (1975)	CS: individual households	A	(−1.00)	
Uri (1975)	TS: monthly aggregate U.S.	A	−0.61	−1.66
FEA (1976)[a]	CS:TS: census regions of U.S., annual	A	−0.19	−1.46
Mount and Chapman (1976)	CS:TS: states, annual	M/A	−0.31	−1.17

Source: Taylor (1977), p. 6.

Notes: CS, cross section; TS, time series; A, ex post average price; M, marginal price; NA, not applicable.

[a]Using 1975 weights for the Northeast census regions.

A.3.4   Fuel Oil*

Robert S. Pindyck (1976)

DATA: Belgium, Canada, France, Italy, the Netherlands,
                         Norway, United Kingdom, United States, and West
                         Germany
                         pre-1974
                         international pooled cross-section time series
METHOD: iterative nonlinear Zellner estimation with country
                         dummy variables
GENERAL MODEL: static version of indirect translog utility function

$$S = f(P_i/M, t)$$

where S = budget shares expended on electricity
   P_i/M = ratio of the countrywide average of the average retail price of
            fuel i (measured in local currency units per teal and con-
            verted to 1970 U.S. dollars) to total residential consumption
            expenditure on energy
      t = time
Results are presented in Table A.11.

QUALIFICATIONS

Static Model. As a static model, lagged responses to a sudden price
change are not incorporated, but rather are assumed to occur in-

**TABLE A.11.   Long-Run Price Elasticities
of Demand for Liquid Fuel (Pindyck's
Model 7)**

Own-price	1962	− 1.46
	1968	− 1.49
Cross-price		
Natural gas	1962	0.08
	1968	0.08
Liquid fuel	1962	0.09
	1968	0.10

Source: Pindyck, 1976, p. 62.

*See also the combined model for natural gas and petroleum products that was developed
by Melvyn Fuss and co-authors, which was reviewed in the section on natural gas (A.3.2).

stantaneously. Therefore, Pindyck's elasticities are assumed to be long run.

Homotheticity. A homotheticity assumption is imposed on the model because Pindyck had difficulty in computing nonhomothetic models (p. 55). Under homotheticity the budget share S is independent of total expenditures M (p. 9). According to Pindyck, the qualification is critical enough to render the results inconclusive.

Capital Stock. The model assumes that the expenditure share on a fuel is independent of the expenditure shares on the energy-using capital stocks.

Data and Results. The quality of the data varies among countries and the results in general are preliminary and incomplete.

INTERPRETATION

Pindyck's own-price elasticity estimate for fuel oil is greater than his estimate for either natural gas or electricity. While recognizing important qualifications to his results, the implication is that consumers respond more to a fuel oil price change than to price changes in natural gas or electricity. As the cross-price elasticity estimates between natural gas and fuel oil and between electricity and fuel oil are relatively small, it can be inferred that fuel oil consumption is altered as a result of own-price changes, while substitution away from fuel oil to natural gas and electricity is minimal. Because of this predicted small tendency toward interfuel substitution as a result of a price increase, aggregate energy consumption is predicted to fall, which is a useful phenomenon when attempting to control or modify total energy consumption. Otherwise, attempts to control consumption of one fuel merely result in increased consumption of another fuel so that, in the aggregate, no progress toward energy conservation has been made.

Pindyck (1976, p. 60) concludes that "much of the increase in oil shares (i.e., budget share expended on fuel oil) could indeed be attributed directly to falling oil prices" over the period 1962–1970. He also acknowledges, however, that "part of this shift in shares was probably due to changing availabilities of fuel-burning equipment." Nevertheless, as mentioned in Sec. A.2 of this report, it is unwise to apply a price elasticity derived from data representing a period of decreasing real energy prices to a future time of increasing real energy prices.

Summary of Recent American Studies

Author	Type of Data	Type of Price	Price Elasticity of Demand for Fuel Oil	
			Short Run	Long Run
Anderson (1973)[a]	TS	A	NA	−1.58
Joskow and Baugh-man (1975)[a]	CS:TS	A	NA	−0.81[b]
Erickson, Spann, and Ciliano (1973)[a]	CS	A	NA	−2.55
Verleger and Sheehan (1974)[b]	CS:TS: states	NA	−0.22	−0.93

[a]Cited in Waverman 1977, p. 7.
[b]Nonlinear equation, elasticity calculated at mean.
[c]Cited in Taylor 1977, p. 25.
Notes: CS, cross section; TS, time series; A, average; NA, not available.

A.3.5   Summary of Canadian Price Elasticities

Energy Form	References	Estimated Price Elasticity	
		Short Run	Long Run
1. Gasoline	Dewees, Hyndman, Waverman (1974)		
	Model (1)	NE	−0.78
	Model (2)	NE	−0.73
	Model (3)	−0.08	−0.40
	Model (5)	NE	−0.69
2. Natural gas	Watkins (1974)		
	Linear model	−0.34 (av.)	−1.83 (av.)
	Double-log model	−0.29 (av.)	NE
	Berndt, Watkins (1977)		
	Model A	−0.14 (av.)	−0.69
	Model B	NE	−0.73
	Fuss, Waverman (1975)		
	Model A		
	Per household	NE	−0.86
	Per capita	NE	−0.96
	Model B		
	Per household	NE	−0.75 (av.)
	Per capita	NE	−0.79 (av.)

Energy Form	References	Estimated Price Elasticity	
		Short Run	Long Run
	Mathewson & Associates (1976)		
	Heating	NE	−1.35 to −1.77
	Appliances	NE	−0.21
	Pindyck (1976)	NE	−1.15
3. Electricity	Fuss, Waverman (1975)		
	Model A		
	Per household	NE	−0.29
	Per capita	NE	−0.34
	Model B		
	Per household	NE	−0.19 (av.)
	Per capita	NE	−0.21 (av.)
	Mathewson & Associates (1976)		
	Model A		
	Heating	NE	−2.28 to −6.07
	Appliances	NE	−0.01
	Model B		
	Heating	NE	−1.20 to −6.86
	Appliances	NE	−0.10
	Pindyck (1976)	NE	−0.81
4. Fuel oil	Pindyck (1976)	NE	−1.49

Note: NE, not estimated.

A.4   CONCLUSIONS

Use of elasticity estimates for the formulation of an energy conservation policy should be only one of several other inputs like consumer surveys and expert opinion. Elasticity estimates should be interpreted with caution and only after consideration has been given to the type of estimation model employed, the other variables employed, the level of aggregation, and the characteristics of the data. Because these factors vary among the studies surveyed, it is possible only to delineate a range within which it is quite probable the "actual" elasticity value lies. Unfortunately, it is impossible to calculate the probability that such an interval contains the "actual" elasticity value.

The ranges for Canadian price elasticities for those studies reviewed are summarized in Table A.12.

As one would expect, the short-run price elasticities for energy are generally lower than the long-run elasticities. The major determinant of this difference is the fact that in the short run, capital stocks are "fixed" and energy demand is therefore relatively unresponsive to changes in

TABLE A.12.   Ranges for Estimated Canadian Price Elasticities of Demand for Energy at the Residential Consumer Level

	Short Run	Long Run
Gasoline	-0.08	-0.40 to -0.78
Natural gas	-0.14 to -0.34	-0.69 to -1.83
(including Mathewson study)		-0.21 to -1.83
Electricity	NA	-0.14 to -0.81
(including Mathewson study)		-0.01 to -6.86
Fuel oil	NA	-1.49

Note: NA, not available.

exogenous variables because of the high costs of switching to a device using a different fuel and the difficulty of fuel substitutability.

In the context of energy conservation, the usual distinction made between inelastic and elastic demand is not relevant per se. The central question is not whether the percentage change in energy consumption is more or less than the percentage change in the energy's price with its resulting impact on revenue, but whether there is a significant induced change in energy consumption at all. The elasticity estimates surveyed imply that the opportunity for induced changes does exist.

A.5   SELECTED REFERENCES ON ELASTICITY

A.5.1   Canadian

Berndt, Ernst R. (1977), "Canadian Energy Demand and Economic Growth," in Campbell Watkins and Michael Walker (eds.), *Oil in the Seventies: Essays on Energy Policy,* The Fraser Institute, Vancouver, B.C., pp. 47–84.

Berndt, Ernst R. (1978), *The Demand for Electricity: Comment and Further Results,* Department of Economics, University of British Columbia, Vancouver, B.C.

Berndt, E. R., and G. C. Watkins (1977), "Demand for Natural Gas: Residential and Commercial Markets in Ontario and British Columbia," *Canadian Journal of Economics,* 10(1), 97–111.

Blomqvist, Ake G., and Walter Haessel (1978), "Small Cars, Large Cars, and the Price of Gasoline," *Canadian Journal of Economics,* 11(3), 470–489.

Denny, M. G. S., and C. Pinto (1975), *The Demand for Energy in Canadian Manufacturing: 1949–70,* mimeographed paper, University of Toronto.

Dewees, D. N., R. M. Hyndman, and L. Waverman (1974), "The Demand for Gasoline in Canada 1956–1972," University of Toronto, Institute for the Quantitative Analysis of Social and Economic Policy, Working Paper No. 7406, June.

Fuss, M. A. (1975), *The Demand for Energy in Canadian Manufacturing: An Example of the Estimation of Production Structures with Many Inputs,* Institute for Policy Analysis, University of Toronto.

Fuss, Melvyn A. (1977), *The Derived Demand for Energy in the Presence of Supply Constraints*, Institute for Policy Analysis, University of Toronto, Working Paper No. 7714, August.

Fuss, M., and L. Waverman (1975), *The Demand for Energy in Canada*, Institute for Policy Analysis, University of Toronto, February, 3 vols., Report Series No. 7.

Fuss, M., R. Hyndman, and L. Waverman (1977), "Residential, Commercial and Industrial Demand for Energy in Canada: Projections to 1985 with Three Alternative Models," in William D. Nordhaus (ed.), *International Studies of the Demand for Energy*, North-Holland, Amsterdam, pp. 151–179.

Gorbet, F. W. (1975), "Energy Demand Projections for Canada: An Integrated Approach," Paper presented at the American Institute of Mining Engineers Meetings, New York, February 16–20 (also available from Energy Mines & Resources Canada).

Hyndman, R., and G. F. Mathewson (1975), *A Report to Ontario Hydro on the Residential Demand for Energy in Canada*, October 15.

Khazzoom, J. Daniel (1973), "An Econometric Model of the Demand for Energy in Canada, Part One: The Industrial Demand for Gas," *The Canadian Journal of Statistics*, 1(1), 69–107.

Khazzoom, J. Daniel (1974), "An Econometric Model of the Demand for Energy in Canada," in Michael S. Macrakis (ed.), *Energy: Demand, Conservation and Institutional Problems*, Massachusetts Institute of Technology, Cambridge, Mass., pp. 360–374.

Khazzoom, J. Daniel (1977), "An Application of the Concepts of Free and Captive Demand to the Estimating and Simulating of Energy Demand in Canada," in William D. Nordhaus (ed.), *International Studies of the Demand for Energy*, North-Holland, Amsterdam, pp. 115–136.

Mathewson, G. F., and Associates (1976), "Residential Demand for Electric Energy and Natural Gas: A General Model Estimated for Canada with Forecasts," Ontario Hydro, *Electricity Costing and Pricing Study*, 4, (October) pp. 1–196.

National Economic Research Associates Inc. (1976) *Analysis of Price Elasticity of Demand for Electricity in Ontario Hydro's Service Area*, report to Ontario Hydro, March.

Pindyck, R. S. (1976), *International Comparisons of the Residential Demand for Energy: A Preliminary Analysis (World Oil Project)*, MIT Energy Lab for National Science Foundation, Cambridge, Mass., September.

Watkins, G. C. (1974), *Canadian Residential and Commercial Demand for Natural Gas*, University of Calgary Discussion Paper No. 30, May.

Waverman, L. (1977), "Estimating the Demand for Energy," *Energy Policy*, (March), 2–11.

A.5.2   American

Anderson, Kent P. (1973), "Residential Demand for Electricity: Econometric Estimates for California and the United States," *The Journal of Business*, 46(4), 526–553.

Balestra, Pietro, and Marc Nerlove (1966), "Pooling Cross Section and Time Series Data in the Estimation of a Dynamic Model: The Demand for Natural Gas," *Econometrica*, 34(3), 585–611.

Baughman, Martin, and Paul Joskow (1975), "The Effects of Fuel Prices on Residential Appliance Choice in the United States," *Land Economics*, 51(1), 41–49.

Berg, Sanford V., and James P. Herden (1976), "Electricity Price Structures: Efficiency, Equity and the Composition of Demand," *Land Economics*, 52(2), 169–178.

Blattenberger, Gail R., Lester D. Taylor and Robert K. Rennhack (1983), "Natural Gas Availability and the Residential Demand for Energy," *The Energy Journal*, 4(1), 23–45.

Chern, W. S. (1978), "Aggregate Demand for Energy in the United States," in G. S. Maddala and others, (eds.), *Econometric Studies in Energy Demand and Supply*, Praeger, New York, pp. 5–43.

Cicchetti, Charles J., and V. Kerry Smith (1975), "Alternative Price Measures and the Residential Demand for Electricity: A Specification Analysis," *Regional Science and Urban Economics*, 5(4), 503–516.

Edmonson, Nathan (1975), "Real Price and the Consumption of Mineral Energy in the United States, 1901–1968," *Journal of Industrial Economics*, 23(3), 161–174.

Grandjean, Burke D., and Patricia A. Taylor (1976), "Public Policy and Renters' Electric Bills," *Social Science Quarterly*, September, 57(2), 437–444.

Halvorsen, Robert (1975), "Residential Demand for Electric Energy," *Review of Economics and Statistics*, 57(1), 12–18.

Hartman, Raymond S. (1979), "Frontiers in Energy Demand Modeling," in J. Hollander, M. Simmons, and D. Wood (eds.), *Annual Review of Energy*, 4, Annual Reviews Inc., Palo Alto, Calif.

Hendricks, Wallace, Roger Koenker, and Robert Pedlasek (1977), "Consumption Patterns for Electricity," *Journal of Econometrics*, 5(2), 135–153.

Houthakker, H. S., and Lester D. Taylor (1970), *Consumer Demand in the United States: Analyses and Projections*, Harvard University Press, Cambridge, Mass.

Houthakker, H. S., P. K. Verleger, Jr., and D. P. Sheehan (1974), "Dynamic Demand Analyses for Gasoline and Residential Electricity," *American Journal of Agricultural Economics*, 56(2), 412–418.

Kraft, John, and Mark Rodekohr (1978), "Regional Demand for Gasoline: A Temporal Cross-Section Specification," *Journal of Regional Science*, 18(1), 45–56.

Lyman, R. Ashley (1978), "Price Elasticities in the Electric Power Industry," *Energy Systems and Policy*, 2(4), 381–406.

Mount, T. D., and others (1974), "Electrical Demand in the United States: An Econometric Analysis," in Michael S. Macrakis (ed.), *Energy: Demand, Conservation, and Institutional Problems*, Massachusetts Institute of Technology, Cambridge, Mass., pp. 318–329.

Nelson, Jon P. (1975), "The Demand for Space Heating Energy," *Review of Economics and Statistics*, 57(4), 508–512.

Peck, A. E., and O. C. Doering III (1976), "Voluntarism and Price Response: Consumer Reaction to the Energy Shortage," *Bell Journal of Economics and Management Science*, 7(1), 287–292.

Ramsey, J., and others (1975), "An Analysis of the Private and Commercial Demand for Gasoline," *Review of Economics and Statistics*, 57(4), 502–507.

Ruffell, R. J. (1978), "Measurement of Own-Price Effects on the Household Demand for Electricity," *Applied Economics*, 10(1), 21–30.

Spurlock, C. W. (1978), "Forecasting Regional Demand for Heating Fuel," *Growth and Change*, 9(2), 29–34.

Taylor, Lester D. (1975), "The Demand for Electricity: A Survey," *Bell Journal of Economics and Management Science*, 6(1), 74–110.

Taylor, Lester D. (1977), "The Demand for Energy: A Survey of Price and Income Elasticities," in William D. Nordhaus (ed.), *International Studies of the Demand for Energy*, North-Holland, Amsterdam, pp. 3–43.

Tyrrell, T. J. (1974), "Projections of Electricity Demand," in Michael S. Macrakis (ed.), *Energy: Demand, Conservation and Institutional Problems*, Massachusetts Institute of Technology, Cambridge, Mass., pp. 342–359.

Wilder, Ronald P., and John F. Willenborg (1975), "Residential Demand for Electricity: A Consumer Panel Approach," *Southern Economic Journal*, 42(2), 212–217.

Wilson, John W. (1971), "Residential Demand for Electricity," *Quarterly Review of Economics and Business*, 11(Spring), 7–22.

Wilson, J. W. (1974), "Electricity Consumption: Supply Requirements, Demand Elasticity and Rate Design," *American Journal of Agricultural Economics*, 56(2), 419–427.

Yang, Yung Y. (1978), "Temporal Stability of Residential Electricity Demand in the United States," *Southern Economic Journal*, 45(1), 107–115.

ILLUSTRATIVE SEMIQUANTITATIVE ENERGY RESPONSE QUESTIONNAIRE

As discussed in the text, it is necessary to gather disaggregated behavioral response data to complement existing econometric studies of price elasticity. This is particularly important in light of the rapidly changing energy supply and demand situation.

Questionnaires should be developed that can assess the responses of individual consumers to a variety of economic incentives and link these potential responses to important consumer attributes. This methodology promises to increase the predictive power of analyses of consumer response, and thereby policy effectiveness.

Initially, such questionnaires should focus on the two largest areas of consumer energy use: private transportation and home heating. They should also incorporate two important types of questions that are sometimes omitted from similar types of research:

1. Questions designed to elicit quantitative (in addition to qualitative) response information.

2. Questions addressing the issue of perceived program effectiveness as well as acceptability. In other words, consumers are asked not only what incentive mechanisms they like, but also what mechanisms they think will influence their behavior, regardless of program acceptability.

Clearly, there are some problems associated with this type of research as consumers are asked to describe their response to hypothetical situations. This consumer information may be distorted by the respondent's ignorance or by an unwillingness to answer correctly.

Nevertheless, these deficiencies do not detract from the value of such a research program as a complement to econometric data and case studies. These questionnaires should be administered to both consumer panels and random public samples, as each category has particular research value. While the use of a consumer panel affords a unique opportunity to assemble a body of useful and accurate consumer attribute information, the well-designed, random public sample guarantees that response data have a high representative and predictive value.

Brief Note on Methodology.   Because of changing interest rates, their prespecification, as in questions 11 and 30, may confound the interpretation of the intensity of selected incentives. There are two possible approaches to this problem:

1. To modify the questions so that discounts from the prime rate are used instead. The principal shortcoming of this alternative is that consumer response may be less clear when the questions are posed in this slightly more abstract fashion.
2. To retain the questions in their current form but adjust for differences from the prime rate after the questions have been answered.

DATE _____

CITY _____

Annual Family Income (Before Taxes)

less than $5000	_____
$5000–$9999	_____
$10,000–$14,999	_____
$15,000–$19,999	_____
$20,000–$25,000	_____
greater than $25,000	_____

Family Size (total number living together, including children, grandparents, etc.)

Please Circle One

1 2 3 4 5 6 7 8 more than 8

Automobile Ownership and Use

1. Do you have a car or truck?

    CAR   TRUCK   MOTORBIKE

    _____

    YES   _____

    NO   _____

    If no, go to question 13

    If yes, how many vehicles do you have?   _____

    make of vehicle(s)   _____

    model year of vehicle(s)   _____

Servicing Your Car

2. How often does your car(s) or truck(s) receive a tune-up?

    three times a year   _____

    twice a year   _____

    once a year   _____

    once every two years   _____

    once every three or more years _____

    other (please specify)   _____

3. Who tunes your car(s) or truck(s)?

    myself or someone else
       providing free labor   _____

    service depot or other place
       that charges for labor   _____

4. Who pays for tune-up?

    myself   _____

    someone else (please specify) _____

5. Suppose the government removed the sales tax on tune-up parts and service for your car. Would you have a tune-up more often?

    YES   _____

    NO   _____

    If yes, how often?   _____

Buying a Car

6. Suppose you are considering the purchase of a car. Would the car's gas mileage be the most important consideration in your purchase decision?

    YES _____

    NO   _____

7. What features are important to you in the purchase of a new car? Please indicate importance by rank (i.e., 1 = most important; 2 = second most important; etc.):

    gas mileage   _____
    size   _____
    purchase cost   _____
    speed   _____
    repair record   _____
    styling   _____
    handling   _____
    comfort   _____
    durability   _____
    other (please specify) _____

8. Would the removal of the sales tax just on cars with very good gas mileage convince you to buy one of these cars?

    YES _____

    NO   _____

9. Suppose the government decided to let you reduce your income taxes if you bought a car with very good gas mileage (e.g., better than 33 MPG, or less than 8.5 liters per 100 km). What is the *minimum* reduction in income taxes that would convince you to buy a new car with very good gas mileage?

    income tax reduced by 5% of cost of car   _____
    income tax reduced by 10% of cost of car   _____
    income tax reduced by 20% of cost of car   _____
    income tax reduced by 30% of cost of car   _____
    income tax reduced by 40% of cost of car   _____
    income tax reduced by 50% of cost of car   _____

10. Suppose the government decided to put an extra tax on cars with poor gas mileage (i.e., "gas guzzlers" that get less than 22 MPG or consume more than 13 liters per 100 kilometers). What level of tax would convince you *not* to buy such a car?

5%   _____
10%   _____
20%   _____
30%   _____

11. Suppose the government decided to grant low-interest loans on the purchase of only those cars with very good gas mileage. What interest rate would convince you to purchase one of these cars with very good gas mileage?

16%   _____
14%   _____
12%   _____
10%   _____
8%   _____
6%   _____
4%   _____
2%   _____

12A. Having considered several policies affecting your purchase decision, how would you rank them in *acceptability* to you? (Please rank by number, where 1 is the most acceptable.)

no sales tax on cars with very good gas
   mileage   _____
extra tax on cars with poor gas mileage   _____
low-interest loans for the purchase of cars
   with very good gas mileage   _____
reduction in income tax for purchase of car
   with very good gas mileage   _____
increase in price of gasoline   _____

12B. Which of these policies would have the *greatest impact* on your decision to purchase a car with good gas mileage?

no sales tax on cars with good gas mileage   _____
extra tax on cars with poor gas mileage   _____
low-interest loans for the purchase of cars
   with very good gas mileage   _____
reduction in income tax for purchase of car
   with very good gas mileage   _____
increase in price of gasoline   _____

Getting to Work

13. Do you commute?

      YES   _____

      NO   _____

  If no, please go to question 22.

13A. When you commute, approximately how many miles is it to your destination?

      fewer than 5 miles (8 km)   _____

      between 5 miles (8 km) and 10 miles (16 km)   _____

      between 10 miles (16 km) and 20 miles (32 km)   _____

      more than 20 miles (32 km)   _____

14. How do you commute?

      (a) alone by car   _____

      (b) with another person by car   _____

      (c) belong to a carpool   _____

      (d) bus/subway   _____

      (e) motorcycle   _____

      (f) bicycle   _____

      (g) walk   _____

      (h) other (specify)   _____

  If you did not answer yes to (a) or (b), please go to question 22.

15. Approximately how much does it cost you to park your car on a typical day? (Circle one.)

      $10 $9 $8 $7 $6 $5 $4 $3 $2 $1   50¢   free

15A. Why don't you use a bus/subway?

      (a) too far to walk to and from terminal   _____

      (b) too expensive   _____

      (c) time-consuming   _____

      (d) bothersome   _____

      (e) uncomfortable   _____

      (f) inconvenient   _____

      (g) not available   _____

      (h) other (please specify)   _____

16. How would you react to each of the following daily charges if they were levied on the driver of all cars carrying no passengers to work?

	50¢	$1.00	$2.00	$4.00	$8.00
(a) Would convince me not to commute in this manner					
(b) Would have no effect on me					

(c) I have no choice _____ (Reason: _____)

17. Which fares (one way) would convince you to take the bus to work?

$1.00	_____
75¢	_____
50¢	_____
25¢	_____
10¢	_____
5¢	_____
free	_____
none of the above	_____

18. How would you react to each of the following percentage discounts on the price of gasoline for all car pool purchases?

	5%	10%	20%	40%
Would convince me to use car pools				
Would have no effect on me				

19. How would you react to each of the following daily car parking rates for people who drive to work?

	$2.00	$4.00	$8.00	$16.00	$32.00
Would convince me not to commute in this manner					
Would have no effect on me					

20. How would you react to each of the following prices of gasoline?

	$1.25/gal (27.5¢/liter)	$1.50/gal (33¢/liter)	$2.00/gal (44¢/liter)	$3.00/gal (66¢/liter)	$5.00/gal ($1.10/liter)
Would convince me not to commute in this manner					
Would have no effect on me					

21A. Having considered several possible policies affecting the use of your car, how would you rank them in terms of *acceptability* to you? 1 = most acceptable; 2 = next most acceptable, etc.

charges for cars with only a driver   _____

cheaper gasoline for car pool use   _____

inexpensive bus/subway fare   _____

expensive parking rates   _____

increases in the price of gasoline   _____

21B. Which of these policies would have the *greatest impact* on your use of your car for commuting?

charges for cars with only a driver   _____

cheaper gasoline for car pools   _____

inexpensive bus fare   _____

expensive parking rates   _____

increases in the price of gasoline   _____

Insulating Your House

22. Do you own a house?

YES _____ NO _____

If no, you have finished the questionnaire.
Thank you for helping us.

22A. Brief description of house:

number of rooms   _____

number of floors   _____

does the house have an attic?   _____

23A. Is your house insulated?

YES _____ NO _____

If no, please go to question 24.

23B. Description of insulation:
      where is house insulated? (e.g., walls, attic, etc.)

      _____

      what is thickness, R value, or type of insulation?

      _____

23C. Who insulated your house?
      you   _____
      previous owner   _____
      builder   _____

23D. When was your house insulated?   _____

23E. What reason or event induced you or the previous owner or builder to insulate the house?

      _____

23F. If the house is already insulated, do you believe or have you been told that it is adequately insulated or well insulated? _____

      If your answer to question 23F was yes, please go to question 35.

24.   Why is your house not (better) insulated?
      insulation cost is too expensive   _____
      savings in fuel bills would be too
         small to bother insulating   _____
      too inconvenient or time-consuming   _____
      other (please specify)   _____

25.   What was your property tax for the most recent collection year?

                                              _____

26.   How many more years do you plan on owning your present house?

                                              _____

27.   How would you react to each of the following *property tax reductions* for one year for insulating your house?

	10% of property tax refunded	25% of property tax refunded	50% of property tax refunded
Would convince me to insulate (or upgrade my insulation)			
Would have no effect on me			

28. How would you react to each of the following reductions in your taxable income for one year for insulating your house?

	$100	$200	$400	$800
Would convince me to insulate (or upgrade my insulation)				
Would have no effect on me				

28A. How would you react to each of the following reductions in your taxes for one year for insulating your house?

	$100	$200	$400	$800
Would convince me to insulate (or upgrade my insulation)				
Would have no effect on me				

29. Suppose it costs $1000 to insulate your house. If loans at very low interest rates were available to help you with the initial cost of purchase, how much of the $1000 could you come up with, assuming that you would like to insulate your house? (Please circle one.)

    $0 $100 $200 $300 $400 $500 $600 $700 $800
    $900 $1000

    If you circled $1000, go to question 31.

30. Considering the size of the loan that you would need to meet the installation cost for insulating your house, how would you react to each of the following annual interest rates?

	16%	14%	12%	10%	8%	6%	4%	2%	0%
Would convince me to insulate (or upgrade my insulation)									
Would have no effect on me									

31. How would you react to each of the following annual taxes which would be levied on owners of houses that are not insulated (or not well insulated)?

	$10	$25	$50	$100	$200
Would convince me to insulate (or upgrade my insulation)					
Would have no effect on me					

32. What type of fuel do you use to heat your house?

     fuel oil ————————

     natural gas ————————

     electricity ————————

     propane ————————

     wood ————————

     other (specify) ————————

33. How would you react to each of the following increases in the price of your home heating fuel?

	+10%	+25%	+50%	double	triple
Would convince me to insulate (or upgrade my insulation)					
Would have no effect on me					

34A. Having considered several possible policies aimed at encouraging the insulation of your house, how would you rank them in terms of *acceptability* to you? (1 = most acceptable, 2 = next most acceptable, etc.)

     property tax reductions ————————

     a reduction in your taxable income ————————

     a reduction in your income tax payable ————————

     low-interest loans ————————

     annual tax for not insulating ————————

     increase in the price of home heating fuel ————————

34B. Which of the following measures is most likely *to convince you* to insulate your house?

     property tax reduction ————————

     a reduction in your taxable income ————————

     a reduction in your income tax payable ————————

     low-interest loans ————————

     annual tax for not insulating ————————

     increase in the price of home heating fuel ————————

35. Which of the following energy-saving features do you have in your house now and which are you considering purchasing?

	Installed now	Considering purchase
Solar space heating		
Solar water heating		
Storm windows		
Storm doors		
Double-glazed windows		
Triple-glazed windows		
Weatherstripping		
Fluorescent lights		
Insulation for hot water heater		
Other (please specify)		

36. Which of the following energy-saving activities do you already undertake?

	Yes	No	Sometimes
(a) turn off the lights when leaving a room			
(b) turn down the thermostat at night			
if yes, by how many degrees			
°F _____ °C _____			
to what temperature?			
°F _____ °C _____			
(c) have the furnace serviced frequently (please specify)			
(d) hang the clothes to dry			
(e) defrost freezer frequently			

Thank You Very Much for Answering This Questionnaire.

COMPILATION OF ENERGY CONSERVATION INCENTIVES

Table 1 is a matrix of possible combinations of incentive measures and targets that policymakers can systematically review to generate alternative program options in the residential sector. Table 2, a listing of the advantages, disadvantages, and major recent references for each option, constitutes a synthesis of the findings relevant to each measure.

TABLE 1

Incentive/Target	Positive Incentives					Negative Incentives				
	Tax Incentives: Credits, Deductions, Rebates, Exemptions A	Grants B	Loans C	Subsidies D	Other E	Taxes F	Fines G	Special Charges or Rates H	Price Increases I	Other J
1. Solar heating	1A	1B	1C	1D	1E	1F	1G	1H	1I	1J
2. Conventional space heating and cooling	2A	2B	2C	2D	2E	2F	2G	2H	2I	2J
3. Hot water heating	3A	3B	3C	3D	3E	3F	3G	3H	3I	3J
4. Swimming pool heating	4A	4B	4C	4D	4E	4F	4G	4H	4I	4J
5. Weatherization	5A	5B	5C	5D	5E	5F	5G	5H	5I	5J
6. Major appliances	6A	6B	6C	6D	6E	6F	6G	6H	6I	6J
7. Urban transportation	7A	7B	7C	7D	7E	7F	7G	7H	7I	7J
8. Intercity transportation	8A	8B	8C	8D	8E	8F	8G	8H	8I	8J
9. Recreational vehicles (boats, planes, RVs, snowmobiles)	9A	9B	9C	9D	9E	9F	9G	9H	9I	9J
10. Unclassified	10A	10B	10C	10D	10E	10F	10G	10H	10I	10J
11. Total residential energy use	11A	11B	11C	11D	11E	11F	11G	11H	11I	11J

TABLE 2

Table Code	Target	Economic Incentive or Disincentive	Special Advantages	Special Constraints or Disadvantages	Comments	References
1A	Solar power	Property tax exemptions for solar installations	Politically popular Easy to administer	Very weak incentive (local tax rate of $10 per $100 on a $6000 solar unit would mean savings of only $150 per year) California experience suggests appraisers do not account for solar power in assessments for taxation	Some feel that inclusion of solar unit in assessed value of house is major deterrent in Canada Could offer rebate on assessed value of solar equipment	Minan and Lawrence, 1978 Berkowitz, 1978
1A	Solar power	Tax deductions on income by solar manufacturers	Would encourage entrants into industry and spread solar technology in Canada Manufacturer, distributor, or contractor would be allowed to deduct fixed amount or percent of income from sale of solar systems; might lower price of systems to end users Administrative mechanism already in place	Definition of solar systems needs to be specified Discriminates against other forms of alternative power systems		Schliflett and Zuckerman, 1978

(continued)

TABLE 2 (Continued)

Table Code	Target	Economic Incentive or Disincentive	Special Advantages	Special Constraints or Disadvantages	Comments	References
1A	Solar water and space heating	Substantial tax credit for solar installations (e.g., >50% of system cost)	Reduces cost of solar systems significantly; speeds adoption. Utilizes tax system rather than creating new bureaucracy. Less political resistance than direct grant or rebate proposal. Promotion of solar energy is politically popular. Solar technology is labor intensive and environmentally sound	Credit claimed by high-income homeowners. Disproportionate use of solar swimming pool heaters when allowed. Reduction in tax revenue. Tax refund does not help initial financing; might couple credit with loan program for maximum effectiveness	Average system cost is approx. $1500. Should certify and test eligible systems in advance. Interest in program increases with time and degree of publicity. Can include other conservation measures under credit terms if installed in conjunction	Foster and Sewell, 1977. Epstein and Barrett, 1977. California Energy Commission, Sept. 1979c
1A	Total residential energy use, but especially solar hot water systems	Tax credit	Strong market impact for solar domestic hot water systems	Credit needs to be about 30–40% of solar costs before any substantial market response is likely. High level of subsidy is politically unacceptable. Tax credits and deductions not appro-	Savings vary with region and are highest for those regions that are coldest and have highest fuel prices. Low-income taxpayers could receive refund where credit exceeds liability	Epstein and Barrett, 1977. DOE AM/IA 1978f. Minan and Lawrence, 1978

1A, B	Total residential energy use, but especially solar hot water systems	Rebates and grants	Because of immediacy of return to purchaser, greater market impact than tax incentive More flexibility and better fit with administrative capabilities	priate for low-income taxpayers Problem of getting political support for substantial subsidy amounts in grant form	Credit should be limited to original investors and not carried forward	Epstein and Barrett, 1977
1B	Solar power	Declining lump sum payments to those installing solar equipment (e.g., $1500 in 1980 declining to a nominal sum in 2000)	Provides most effective incentive to "early adopters" Helps ease high initial cost of installation Declining grants in future would reflect economies of scale Incentive measure is automatically phased out	Payments will likely go to those with high income who tend to be early adopters Must specify solar measures or certify units; care must be taken not to discriminate against home-built units		EM&R, 1978b
1B	Increased solar R&D	Grants for solar research, development and demonstration projects	Aids commercialization of solar technology Can give exposure to many low-cost projects; program should aim for high visibility Promoting solar technology is popular political stance	Should be coordinated with federal efforts	Should draw on university expertise Solar special interest groups are helpful in project screening Passive solar systems should not be overlooked	SERI, Jan. 1979a

(continued)

TABLE 2 (Continued)

Table Code	Target	Economic Incentive or Disincentive	Special Advantages	Special Constraints or Disadvantages	Comments	References
IC	Total residential energy use, but especially solar hot water systems	Loans	Strong impact on combined heating and hot water systems in new construction where loan size is large enough to bear the transaction costs involved	Only marginal increase expected in use of solar systems Full exposure of purchaser Institutional and administrative obstacles Loans are associated with low-income status and major paperwork requirements Government must assume the administrative responsibility		Epstein and Barrett, 1977
1E	Total energy use in the home (esp. home heating)	Special home insurance rates for solar energy use			Subsidy of additional premiums by federal government if necessary to overcome industry conservatism	Foster and Sewell, 1977
1E	Industry incentives to solar power	Loans to solar equipment manufacturers for R&D and marketing (particularly to develop standardized units)	A "push" strategy; advantages are that manufacturers are fewer in number than consumers Loans to consumers are "pull" strategy, creating demand Type and rate of development at government's control	Many programs in effect for R&D, few for marketing and promotion		Schliflett and Zuckerman, 1978

1E	Solar power	Special package of in- centives to multiple dwelling owners: (1) accelerated de- preciation; (2) de- ductibility of main- tenance expenses; and (3) $1500 lump sum initial payment	heating systems are 75% of cost of sin- gle-family unit; Owners are typically corporations who respond to ac- celerated deprecia- tion costs; Solar most econom- ical for multiple family units	for utility bills for incentive to work; conflicts with change away from master metering	
1E	Solar power	Federal insurance of hazards associated with installation and use of solar equipment (e.g., wind, hail, trees, and vandalism)	Hazards constitute major impediment to adoption; Would overcome high cost of insurance	Would require admin- istrative apparatus or contracting through private agency with experi- ence in solar field; Initial cost of solar in- stallation is greatest impediment; insur- ance availability is secondary consider- ation	Schliflett and Zucker- man, 1978
1E	Solar power	Utility ownership and leaseback of solar systems	Would overcome ini- tial high cost of solar systems; Utility reliability and familiarity with lo- cal conditions would make pro- gram attractive; could be supple- mented by special rates for backup systems; Utilities would help participate in devel- opment of solar technology if per- ceived as profitable opportunity	Utilities would incur costs of training in- stallers, repairmen, etc., which would raise costs to all users; Potential utility mo- nopoly on solar equipment	Schliflett and Zucker- man, 1978

(continued)

TABLE 2 (Continued)

Table Code	Target	Economic Incentive or Disincentive	Special Advantages	Special Constraints or Disadvantages	Comments	References
2A	Total residential energy use, but esp. space heating	Tax credits for persons renting or buying new high-density dwellings located in energy-efficient areas	Would stimulate construction of energy-efficient residential development; If applied to central city locations would reduce commuter's traffic, pollution, and parking problems	Program indirect; May need high credits to get response; Government loses revenue; May be complicated for people to determine eligibility	Limit applicability to solar heated/cooled structures; Limit to central city high-density housing; Formulate building standards; minimum number of units	Portland V. 3B, 1977d
2A	Purchase of air conditioners	Adjust sales tax up for low-efficiency units and down for high-efficiency units	Visible cash incentive at time of purchase	Inequity between low- and high-income purchasers; Paperwork for retailers; Consumer awareness may be lacking before the purchase decision has been made		Arthur D. Little, 1975
2A	Residential space heating	New residences: (a) conservation tax rebates or credits (e.g., $200), (b) tax rebates on 2-story homes (e.g., $200)	Expected reduction in new residential sector energy use in Portland: (a) 14.8%, (b) 1.8%	No immediate effects; Less efficient than building standards alternative		Portland V. 3, 1977b
2A	Home heating improvements	Investment tax credits	Lowers the cost of investing in energy conservation	Program is indirect and may require large credits to create an adequate incentive		Portland V. 3A, 1977c

				Disadvantages	References
2A	Space heating	Tax incentives to municipalities to promote district heating		Government loses revenue. May be complicated for residents to interpret	Knelman, 1975; Berg, 1973
2B, C	Home heating improvements	Government low-interest loans or direct grants	Assistance to needy	Some technological disadvantages (see Berg)	Portland V. 3A, 1977c
2C	Total residential energy use, but esp. solar space heating	Low-interest loans for individual buildings, esp. (a) high-density housing in energy efficient city areas, (b) new rental units in existing homes, (c) new 2-story single-family homes	Reduce high first cost of solar heating equipment, and would stimulate construction of energy efficient residential development. Accelerates change-over to higher-density housing	Bureaucracy required to administer program. Large portion of housing. Program indirect. May need large loans to get significant response. Government loses revenue. Bureaucracy required. May be complicated for people to determine their eligibility	Portland V. 3B, 1977d; Foster and Sewell, 1977
2E	Total energy use in residences	Incentives for total energy systems (e.g., small complex of buildings supplied with electricity and reject heat) = 100–300 residences	Avoid problems of large-scale centralization, e.g., high-temperature combustion pollutants (NO_x), large capital costs of construction and distribution network	Total energy plants, because they are small, have a lower effective temperature of combustion and are consequently less efficient. Still dependent on fossil fuels. Generally involve high maintenance requirements	Berg, 1973

(continued)

TABLE 2 (Continued)

Table Code	Target	Economic Incentive or Disincentive	Special Advantages	Special Constraints or Disadvantages	Comments	References
2F	Purchasers of air conditioners	Annual excise tax on units that exceed a stated level of electricity consumption		Cost of collection Owner identification problems Political acceptability Not visible at point and time of sale		Arthur D. Little, 1975
5A	Thermal integrity of existing residential structures	Income tax exemptions or income tax credits	Not a direct cost to government, but a reduction in revenue Same percentage for all income groups	If payback period is too long, owners may not wish to participate Generally does not provide an incentive for rental housing	Increasing income tax exemption to greater than 100% of original cost aids incentive Must address issue of rental housing	Dole, 1975
5A	Weatherization	Residential energy conservation tax credit	Tax credit used for cost-effective measures taken by consumer to reduce energy consumption Many devices together work better than single measure (insulation)	Care must be taken that cost of subsidy to government does not exceed value of energy saved Might encourage indiscriminate or uninformed expenditures	Use income eligibility, so only those of low income qualify	Hyatt, 1977
5A	Weatherization	Tax credit for home weatherization	Weatherization, particularly ceiling insulation, is the most cost-effective conservation measure	Credit does not insure that homeowner's actions are cost effective Poor have badly insulated homes and benefit least from tax credit Program itself cannot account for effec-		U.S. Community Services Administration, April 1977a

(continued)

	Policy	Description	Advantages	Disadvantages	Comments	Reference
				Tax credits are "after-the-fact" reimbursement and do not help with initial investment which the poor cannot afford. Probable disproportionate distribution of benefits to high-income taxpayers		
5B	Low-income weatherization	Direct grants to low-income homeowners for weatherization	Targets weatherization for those most in need. Some advantages as public employment program. Tends to focus on elderly poor	Fuel savings less than expected because low-income families now living in discomfort choose higher comfort levels with increased energy use. Difficult to control for "optimum" expenditure per house. Does not touch low-income people in rental housing. Difficult to verify cost expenditures	Should be compared to alternative income transfer programs for poor. Program requires "asset" or "means" test. More materials produce better results. Results must be inspected for quality	U.S. Community Services Administration, April 1977a
5C	Low-income weatherization	Loan program for weatherization	Reaches low income and elderly people directly. Benefits of program increase as energy cost rises. In inflationary times, recipients have added benefit of repaying fewer real dollars	Canada already has weatherization program in effect	Studies indicate that program should maximize available cash flow rather than minimize payback period. Program could be directed at other weatherization methods (caulking, weatherstripping, etc.) to complement insulation	U.S. C.S.A., 1977b

TABLE 2 (Continued)

Table Code	Target	Economic Incentive or Disincentive	Special Advantages	Special Constraints or Disadvantages	Comments	References
5C	Weatherization	Utility-sponsored home weatherization loans	Closer contact and knowledge of local housing conditions Able to target areas of special need Successful conservation programs may delay new generation capacity Homeowners trust utilities' recommendations Program promotion facilitated through bill stuffers	Utilities will not pursue independently without government directive Utilities tended to charge maximum permissible interest (utilities have high costs of capital) Cost of program compensated for by general rate increase	Utilities uncertain whether conservation programs are in their interest Program must include full energy audit and postinstallation inspection Can use aerial infrared sensing to see which homes require it No downpayment is great advantage	Booz-Allen, June, 1978
5C	Weatherization	Home improvement loans tied to current government-funded or assisted housing	Much building done in public sector where government can set terms		Alternative is federal efficiency standards	Berg, 1973
5C	Weatherization	Utility company or bank loans	Many of funded conservation actions are immediately cost effective Utilities and financing institutions are familiar and credible sponsors to homeowners	Many only elicit a small participation rate	Utility should offer installation and inspection as well as financing	Portland V. 3A, 1977c
5C	Weatherization	Government loan guarantees and interest reductions	Would help those unable to qualify for financing	Not deemed to be effective: (a) homeowners would		Dole, 1975

Code	Category	Instrument				References
			Benefits accrue to society despite loan defaults	probably already qualify for home improvement loan without government guarantee; (b) guarantee might lower the interest costs, but probably not more than 1%; (c) people may apply for low interest loans and use them for other purposes		Dole, 1975
5D	Weatherization	Interest subsidy program		Not a major incentive because only a small difference in monthly payments		
6A 6F	Use of energy-intensive appliances	Tax on purchase (e.g., hot water heaters, electric home appliances, air conditioner, furnace), or graduated taxes that vary inversely with energy efficiency	Avoids perceived unfair burden on previous purchase of wrong equipment	Does not affect hours of use (i.e., no incentive for efficient use) Little short term effect because of emplaced equipment	Detailed classification schedule required, e.g., one rate for furnaces sold in cold climates, another rate for warmer climates	Seidel, 1973; Brannon, 1974; Dole, 1975
7A	Reduce energy use in automobile sector	Income tax credits for vanpool investments and accelerated depreciation	Should be used where goal exceeds levels attainable by voluntary program Easy to administer through tax codes	Must overcome obstacles of insurance liability Working hours may need adjustments Requires employee time to set up matching	Can be accompanied by preferential parking for additional effect Government sector and large employers are best targets Discourage current employer subsidization of employee parking	EPA, Denver, Mar. 1978 U.S. FEA. 1976, Economic Impact Analysis of Proposed Energy Conservation Contingency Plan

(continued)

TABLE 2 (Continued)

Table Code	Target	Economic Incentive or Disincentive	Special Advantages	Special Constraints or Disadvantages	Comments	References
7A	Gasoline use	Incentives for carpooling and vanpooling, e.g., tax credits for purchase of vehicles and removal of excessive insurance requirements		Switch from public transit to carpools will increase vehicle miles traveled. Could be considered politically as subsidy to labor	People who carpool cite parking savings as primary reason. For best results, government actions must be aimed at employers rather than poolers. Most effective in medium to large urban areas	Portland V. 3B, 1977d; Voorhees, 1976; FEA, Mar. 1976c
7A	Increased vanpooling	Allowing investment credit plus special "energy investment credit" to businesses purchasing vans for employee vanpooling	Would encourage vanpooling by large employers. Very small loss of revenue	Negligible energy savings would result because of limited applicability		U.S. Senate Report 95-529 (1977)
7D	Urban transportation, commuting	Employer subsidization or public subsidization of transit fares by tax deductibility	Shown to be effective incentive (Portland). Eliminate employee's commuting expenses. Program could subsidize all or a portion of bus fares	Unfairness to employees unable to take advantage of the incentive. Substantial employer expense	Subsidy could apply to any nonautomobile method of commuting. Employers should be encouraged to discontinue paying for employee parking	EPA, Colorado, Mar. 1978
7F	Gasoline use	Parking tax on parking lots and parking meters (time dependent)	Income generation scheme for cities as well (San Francisco generated $5.5 million in 1 yr). Commuters more sensitive than shoppers	Resistance by urban merchants. Must simultaneously provide improved urban transportation	Exempt carpoolers and vanpoolers. Can be flat rate or percentage. Should be municipal responsibility	Brannon, 1974; FEA, 1976c

		Advantages	Considerations	Reference
7H	Reduce energy use in automobile sector	Differential and progressive parking rate structure/area of facility tolls	Easy to implement, visible to user at time of use Incentive to multioccupancy vehicles Selectivity	Merchant resistance Enforcement problems with "area" passes or need for supplemental downtown license	Voorhees, 1976
7H	Urban transportation, Commuting	Graduated fee parking meters and lots	Increased fees for all-day commuters Free or reduced-cost parking near mass-transit terminals Local solution is less noticeable Less regressive than gasoline tax; raises cost of driving so effects similar	Must have adequate mass transit in place as alternative for commuters	Brannon, 1974
7J	Reduce energy use in automobile sector	Elimination of employee parking subsidies	Easy to implement, visible to user at time of use	Must compensate for loss of this benefit	Voorhees, 1976
8A	Intercity passenger travel	Refundable tax credit to bus operators for terminal renovation and fare reductions	Bus travel is most economical Credit earmarked for terminal renovations (and fare reductions) to increase ridership Credit can be calculated by number of reduced fares allowed Would help prevent decline in ridership as well Savings in fuel substantial	Must monitor for base fares against which reductions apply Cannot compute extent of response accurately; increased ridership based on many noneconomic factors Charter buses might be included	U.S. Senate Report 95-529, 1977

(continued)

TABLE 2 (Continued)

Table Code	Target	Economic Incentive or Disincentive	Special Advantages	Special Constraints or Disadvantages	Comments	References
8D	Intercity passenger travel	Subsidize passenger rail and bus traffic	Would promote attractive alternative to air travel	Study found disappointing energy savings—(4%) resulting after 20% decrease in fares Few air travelers shifted to bus Speed of mode and quality of terminal facilities often more important than fare	Subsidization of rail more promising than of buses	FEA, April 1977e
8F	Air travel	Tax on aviation fuel	Air travel (most energy-consuming mode) Not regressive in the sense that income and air travel are related		Cannot significantly reduce transportation energy consumption without air transportation policy	Brannon, 1975
8F, I	Reduction in intercity air passenger travel	Increased air fares through price increases or taxes	Very effective in producing energy savings Air travelers are most sensitive to air fare increases; coupled with gasoline price increases, very large savings result Major reductions in energy consump-	Produces substantial deterioration in system quality (travelers drop out of system entirely) Reduction in air travel will not necessarily lead to increased bus and rail traffic	Alternative approach is reduction in availability of aviation fuel to percentage of baseline Policies encouraging higher airline load offer greatest potential Energy use related to speed, which is re-	FEA, April 1977e

		tion are possible only by making major air mode changes	lated to time, so reduced time results in inefficiency; if time is an issue, passengers won't switch modes	
7 & 8A	Automobile use	Would not require investment in more fuel-efficient automobile Would promote development of such devices (mpg conversion, enhancement equipment) Would affect current stock rather than future stock	Effectiveness of devices uncertain Devices exist at present	Gallagher, 1975–76
		Tax deduction of fuel-conserving devices on automobiles		
7 & 8F	Automobile use	Higher fixed tax on new car sales	Little promise of gasoline conservation in near term Does not affect existing stock Adverse employment effects	Wildhorn, 1976; Taylor, 1975
7 & 8F	Automobile use	Annual tax on automobile ownership (new and used cars) or vehicle characteristics	Complete coverage of automobile stock Annual tax would accelerate rate at which older cars were scrapped Penalizes owners of existing stock Question of which level of government administers	Brannon, 1975

(continued)

TABLE 2 (Continued)

Table Code	Target	Economic Incentive or Disincentive	Special Advantages	Special Constraints or Disadvantages	Comments	References
7 & 8F	Automobile use	Gasoline tax	Applies to all cars on the road Only way of achieving significant short-term results (e.g., 1–4 yrs) (Note: Difference of opinion over extent of short-run effects; see Brannon, 1974) Most of expected impact occurs as a reduction in vehicle miles traveled, i.e., elimination of trips Gives incentive to alter composition of car stock Estimated effects: (U.S.) If +15–45¢/gal, then 16–41% reduction in gasoline consumption expected by 1980 Increased revenue could be used for public transit Specific tax on gasoline most effective in reducing gasoline demand without raising all energy sector prices	Does not eliminate itself when no longer necessary Impact of tax declines over time because of rising income and inflation Reduction in personal mobility If supply is not perfectly elastic, part of the tax will be borne by producers and not passed on to consumers Employment decrease in the auto-related sector Does not lead to significant increases in car efficiency Question of regressive nature is contentious (see Taylor, p. 73) Regressive Potentially low short-run elasticity because of existing stock of cars Short-term effect only Need uniform application to prevent avoidance	Longer-run mandating of technological change may be necessary (i.e., a noneconomic mechanism, e.g., mandated weight changes, new car fuel economy, radial tires, aerodynamic body design, continuously variable transmission, supercharged stratified rotary spark ignition engine) Exemption of certain gasoline uses from tax (but potential black market in two-price system) Regressivity can be overcome by tax rebates	Wildhorn, 1976; Taylor, 1975; Brannon, 1974; Portland V. 3B, 1977d

			Can help secure energy independence	High-income people may not respond; Serious political opposition; Use of other fuels might rise under this tax; Would not necessarily combat automobile inefficiency	Gallagher, 1975–76; DOE EIA/0102/3, 1978e
7 & 8F	Automobile use	Progressive tax on the shortfall of fuel economy from a specified standard. "Gas guzzler tax" increases with each mpg by which model falls below the year's mileage standards	Promotes automobile efficiency; Places cost on firm and individual rather than society; Less regressive than gas tax; Tax payable by manufacturers; Ease of administration; Tax directly aimed at petroleum consumption; Attractive to consumers because savings are immediate; More efficient than regulation	Deviations between actual on-road fuel economy and standards; standards are an optimum, so savings of petroleum are overstated; Potential negative employment effects as car sales decline; Some evidence of increased driving in higher efficiency automobiles offsetting savings; Allows inefficiency to continue if paid for	Alternative approach is to prohibit sales of automobiles below a given efficiency (approach preferred by public); Standards must be set high enough to compensate for test/on-road deviation; To offset decline in large-car sales, offer rebates on smaller cars
7 & 8F	Automobile use	Gas tax on new car purchases (based on amount of gas to be consumed over life of car)	Similar to graduated tax on vehicle mpg; Tax could be collected immediately for alternative use; Affects choice of stock; Larger initial tax has more impact than small tax per gallon	1% tax on vehicle consuming 20 mpg would yield only $35 (assuming 100,000 mi life and 9% discount); Tax would have to be substantial to have effect	Brannon, 1974

(continued)

291

TABLE 2 (Continued)

Table Code	Target	Economic Incentive or Disincentive	Special Advantages	Special Constraints or Disadvantages	Comments	References
7 & 8H	Automobile use	Graduated auto registration fees based on energy efficiency, e.g., weight or engine displacement	Incentive to purchase energy-efficient cars Estimated reduction in transportation sector energy use in Portland by 1995: 3.4% Gains in real income could offset vehicle taxes	Regressive High-income people may not respond Possible adverse employment effects	$500 would be minimum effective tax to induce a shift to smaller cars Public prefers restricting weight and horsepower to tax	Portland V. 3 & V. 3B, 1977b,d; FEA, Mar. 1976c
7 & 8H	Automobile use	Highway tolls (e.g., congestion and peak-period tolls)	Marginal-cost approach, therefore an efficient measure for controlling peak-period congestion and consequent excessive gasoline use	Social/political resistance	Area tolls are another alternative that might not apply for high-occupancy vehicles	Brannon, 1974; Elliot, 1975
7 & 8H	Reduce energy use in automobile sector	Multiple-car-ownership tax surcharge	Should be used where goal exceeds levels attainable by voluntary program			Vorhees, 1976
10F	Use of specific fuels	Additional tax on electricity, natural gas, fuel oil (e.g., 25%)	Potential gain: 7% across the board reduction in energy use in the residential sector Shift usage away from fuel oil where alternatives exist	Political problems with acceptance		Dole, 1975

11D	Direct energy subsidy to poor (energy stamps)	Would offset regressive impact of rising energy prices	Difficult to administer Might encourage greater fuel consumption Should not be fuel-specific or switching will occur	Generally subsidizing one form of consumption is inefficient as means of income redistribution	Aman and Howard, 1977
11F	Total energy use	Such taxes will reflect social costs associated with resource usage Environment will be cleaner by amount by which firms clean up plus foregone production Consumption will shift to less polluting alternatives	Prices of energy sources causing negative environmental effects will rise and contribute to inflation Energy production may be curtailed	Types of costs to be internalized include oil spills, strip mining reclamation, nuclear risk, etc.	EPA R5-73-021, 1973
11F	All consumer energy use	Tax on all energy (on Btu value) at retail level Raises price of energy inputs; leads to substitution of labor and capital Raises significant revenue for other uses	Ignores fuel quality and availability Regressive High administration and compliance costs (as compared to the use of this tax at the manufacturer's level) Tax is too broad (e.g., penalizes energy use in desirable activities such as mass transit) Tax has no effect on supply Must be permanent if independence is to be achieved Ignores intermediate demand for energy	Needs income redistribution measures, e.g., increased welfare payments, negative income tax, refundable tax credit, exemption for low-income families	Brannon, 1974; Brannon, 1975; Seidel, 1973; National Science Foundation, Mar. 1975

TABLE 2 (Continued)

Table Code	Target	Economic Incentive or Disincentive	Special Advantages	Special Constraints or Disadvantages	Comments	References
11F	Reduced energy use for home	Tax on electricity, natural gas, and home heating oil use	Generates tax revenue. Relatively simple to administer	Need high tax to elicit good response. Regressive. Needs uniformity across fuels to prevent undesirable switching	Income redistributional measures	Portland V. 3A, 1977c; Lehman and Warren, 1978
11H	Electricity consumption	Flat rates	Rates for small residential users would decrease; bills for larger customers would increase (equity problem). Would promote decline in electricity consumption and shift to oil and gas in long run	Rate not cost based. Not cost based and offer no incentive for load control (elimination of system peaking). Net effect on capacity is uncertain. Canada has no shortage of electricity yet		Nemetz, 1979, and this volume
11H	Electricity consumption	Inverted rates	Small users will enjoy cost reduction. Capacity requirements should decrease; total consumption will decline	Rate not cost based. It is growth in peak demand that determines need for new capacity. Large users subsidize small users (eventually leave system). Increased use of oil and gas	In empirical tests in Vermont, customers' bills rose	Nemetz, 1979, and this volume
11H	Electricity consumption	Penalty pricing. Not a rate, but excess charge when de-	Addresses problem of reducing peak-period demand and	For rate to be effective, customer must know when de-	A preset circuit breaker functions in same manner	Nemetz, 1979, and this volume

	mand (kW) exceeds a certain level		thus limits growth in capacity	mand is approaching limit; requires alarm	Use of demand "limiters" in Europe	
11H	Electricity use	Peak-load pricing	U.S. experiments have shown significant and favorable response to time-of-use rates	Traditional requirements for fair return on utility capital; Political practicability (e.g., change in the status quo); Uncertainty about exact response	Exemption for low income families to ease adverse effects; Requires consumer education program	Doctor, 1972; Berman, 1972; Ekholm, 1977; DOE/ERA-0011 1978c; and this volume
11H	Electricity use	Peak-load pricing, esp. time-of-day tariffs	Slows need for new generating capacity, beneficial environmental effects	If defined peak is too narrow, may induce a shift to an adjacent period and create a new peak; If peak period is defined too long in rate structure, may inhibit shifting; Not cost-free, e.g., administration and metering costs (a benefit-cost problem)	Potential introduction for larger customers only; Try programs on voluntary basis	Manning et al, 1979; Acton, 1978b; DOE/ERA-0011 1978c; and this volume
11H	Electricity consumption	Ratchet rates (surcharge levied per kWh when customer's demand exceeds a certain level)	Rate is simple to administer and understand; Rates can also be a surcharge over previous year's billing; offers flexibility and tailors incentive to each household; No additional metering costs; Easy transition from old rate			Nemetz, 1979, and this volume

(continued)

TABLE 2 (Continued)

Table Code	Target	Economic Incentive or Disincentive	Special Advantages	Special Constraints or Disadvantages	Comments	References
11H	Electricity use	Peak-load pricing, esp. seasonal tariffs	No important costs of converting to such a rate. More accurately reflects costs of hydro utilities	May not be any significant effects of raising the price per kWh for several months		Manning et al, 1979; Acton, 1978b; DOE/ERA-0011 1978c; and this volume
11I	Natural gas use in residence	Price increase	In theory, will affect two principal components: (i) heating use (= 70%), (ii) nonheating use (= 30%)	Empirical findings: relative insensitivity to price for heating component. Posited explanations: (a) lack of other fuels at competitive prices, (b) relative affluence of the average gas-consuming customer		Lehman and Warren, 1978
11J	All residential energy use	An increase in energy prices without any other government actions	Hypothesized U.S. results: If fuel cost in 2000 A.D. is 140% of 1976 in real terms, then it is expected that there would be a 14% cut in energy use *ceteris paribus* (i.e., the growth rate in energy use would be reduced to 1.7% per year)			Hirst and Carney, 1978

			Pros	Cons	Other	Reference
11I	Total energy use	Price increase as a result of domestic decontrol of oil prices	Encourages conservation Stimulates supply Decreases oil imports Improves balance of payments Eventual decontrol must be achieved Savings resulting from discontinuance of oil subsidy	Increased inflation Highly regressive (low-income families unable to respond by buying more efficient stock) Would involve windfall profits for oil companies Consumer response unknown (whether increased conservation, decreased saving, or increased spending on more efficient stock) Regional disparities in West because of longer travel distances	Could undertake methods to assist low-income families Tax oil companies and use proceeds for mass transit Eventual effects of phased and sudden decontrol would be same	U.S. Congress Background paper, *The Decontrol of Domestic Oil Prices*, May 1979
11J	All consumer energy use	Individual energy tax return	Permits introduction of a graduated rate schedule	Complicates calculation and auditing Helps poor		Brannon, 1974
11J	Residential energy use	Penalties imposed at the extreme end of the average computed electricity consumption over the last 3 years	Gradual reduction in energy use, e.g., possible phase in: 1st-year penalty—anything exceeding past 3 years' average consumption; 2nd-year penalty—anything exceeding 95% of past 3 years' average, etc.			Portland V. 3, 1977b

(continued)

TABLE 2 (Continued)

Table Code	Target	Economic Incentive or Disincentive	Special Advantages	Special Constraints or Disadvantages	Comments	References
11J	Total use of energy over the lifetime of residence	Formal recognition of life cycle costing in home energy use	Reduces energy waste over longer term	Economic/institutional barriers to adoption, e.g., potentially higher initial capital cost of purchase	Range of possible government incentives to reduce perceived front-end costs of energy-efficient residences. Note: several U.S. states require that life cycle cost estimates be used in government building construction	U.S. ERDA, 1976
11J	Total residential energy use	Energy use budgets, e.g., government establishes maximum amount of energy that should be used in each household per unit of time—then penalties, fines or surcharges are levied for excess use		To be equitable, account must be taken of number of people in family, age and health, house size, special facilities, local climate, past energy use. High government administration costs		Dole, 1975

REFERENCES

Acton, Jan Paul (1980), *"Electricity Prices and the Poor:* What Are the Effects and What Can We Do?" In *Report from Symposium on Energy Pricing and the Poor,* National Economic Development and Law Center and California Research, Berkeley, California, February, pp. 36–44.

Acton, Jan Paul, and R. S. Mowill (1975), *Conserving Electricity by Ordinance: A Statistical Analysis,* Rand Corporation (R-1650-FEA), Santa Monica, Calif., February.

Acton, Jan Paul, and Ragnhild Mowill (1976), "Regulatory Rationing of Electricity Under a Supply Curtailment," *Land Economics,* Vol. 52, No. 4, November, pp. 493–508.

Acton, Jan Paul, B. Mitchell, and R. Mowill (1976), *Residential Demand for Electricity in Los Angeles: An Econometric Study of Disaggregated Data,* Rand Corporation (R-1899-NSF), Santa Monica, Calif., September.

Acton, Jan Paul, B. M. Mitchell, and W. G. Manning (1978a), *Projected Nationwide Energy and Capacity Savings from Peak-Load Pricing of Electricity in the Industrial Sector,* Rand Corporation (R-2179-DOE), Santa Monica, Calif., June.

Acton, Jan Paul, W. G., Manning, and B. M. Mitchell (1978b), *Lessons to Be Learned from the Los Angeles Rate Experiment in Electricity,* Rand Corporation (R-2113-DWP), Santa Monica, Calif., July. Also as "Lessons from the Los Angeles Rate Experiment in Electricity," in John L. O'Donnell (ed.), *Adapting Regulation to Shortages, Curtailment and Inflation,* Michigan State University Public Utilities Studies, East Lansing, Michigan, 1977.

Acton, Jan Paul, B. Mitchell, and W. Manning (1978c), "European Industrial Response to Peak-Load Pricing of Electricity, with Implications for U.S. Energy Policy," in Canadian Electrical Association, *Marginal Costing and Pricing of Electrical Energy, A State-of-the-Art Conference,* May, pp. 248–267.

Acton, Jan Paul, B. M. Mitchell, and W. G. Manning (1979), "Peak Load Pricing of Electricity," in P. N. Nemetz (ed.), *Energy Policy: The Global Challenge,* Institute for Research on Public Policy, Montreal.

Aman, Alfred C., Jr., and Glen S. Howard (1977), "Natural Gas and Electric Utility Rate Reform: Taxation Through Ratemaking?" *The Hastings Law Journal*, **28** (May).

Anderson, D. (1972), "The Demand for Gasoline," U.S. Department of Transportation, Transportation Systems Center, Cambridge, Mass.

Anderson, Kent P. (1973), "Residential Demand for Electricity: Econometric Estimates for California and the United States," *The Journal of Business*, **46**(4), 526–553.

Arkansas, Public Service Commission (1979?), *Final Report, Arkansas Demand Management Demonstration Study*.

Atkinson, S. E., and R. Halvorsen (1980), "Automatic Price Adjustment Clauses and Input Choice in Regulated Utilities," in Peter N. Nemetz (ed.), *Energy Crisis: Policy Response*, Montreal: Institute for Research on Public Policy, pp. 185–196.

Averch, J., and L. L. Johnson (1962), "Behavior of the Firm Under Regulatory Constraint," *American Economic Review*, **52** (December), 1053–1069.

Balestra, Pietro, and Marc Nerlove (1966), "Pooling Cross Section and Time Series Data in the Estimation of a Dynamic Model: The Demand for Natural Gas," *Econometrica*, **34**(3), 585–611.

Baughman, Martin, and Paul Joskow (1975), "The Effects of Fuel Prices on Residential Appliance Choice in the United States," *Land Economics*, **51**(1), 41–49.

Baumol, William J., and Wallace E. Oates (1975), *The Theory of Environmental Policy*, Prentice-Hall, Englewood Cliffs, N.J.

Berg, Charles A. (1973), "Energy Conservation Through Effective Utilization," *Science*, July 13.

Berg, Sanford V., and James P. Herden (1976), "Electricity Price Structures: Efficiency, Equity and the Composition of Demand," *Land Economics*, **52**(2), 169–178.

Berkowitz, M. K., (1978), "Incentive Schemes for Encouraging Solar Heating Applications in Canada," in Peter N. Nemetz (ed.), *Energy Policy: The Global Challenge*, Institute for Research on Public Policy, Montreal.

Berman, M. B., et al. (1972), *The Impact of Electricity Price Increases on Income Groups: Western United States and California*, Rand Corporation, Santa Monica, Calif.

Berndt, Ernst R. (1977), "Canadian Energy Demand and Economic Growth," in Campbell Watkins and Michael Walker (eds.), *Oil in the Seventies: Essays on Energy Policy*, The Fraser Institute, Vancouver, B.C., pp. 47–84.

Berndt, Ernst R. (1978), *The Demand for Electricity: Comment and Further Results*, Department of Economics, University of British Columbia, Vancouver, B.C.

Berndt, E. R. and G. C. Watkins (1977), "Demand for Natural Gas: Residential and Commercial Markets in Ontario and British Columbia," *Canadian Journal of Economics*, **10**(1), 97–111.

Berndt, E., and D. Wood (1975), "Technology, Prices and the Derived Demand for Electricity," *The Review of Economics and Statistics*, **57** (August), 259–268.

Blomqvist, Ake G., and Walter Haessel (1978), "Small Cars, Large Cars, and the Price of Gasoline," *Canadian Journal of Economics*, **11**(3), 470–489.

Bonbright, James C. (1961), *Principles of Public Utility Rates*, Columbia University Press, New York.

Booz-Allen and Hamilton, Inc. (1978), *Utility Sponsored Home Insulation Programs*, report prepared for U.S. Department of Energy, June.

Bradt, Ronald D. (1983), "Annual Report for 1982 on the Renewable Energy Income Tax Credit and Direct Refund Program for the State of Wisconsin," DILHR, June 10.

Brannon, Gerard M. (1974), *Energy Taxes and Subsidies*, Ballinger, Cambridge, Mass.

Brannon, Gerard M. (ed.) (1975), *Studies in Energy Tax Policy*, Ford Foundation, Ballinger, Cambridge, Mass.

Brookhaven National Laboratory (1977), *Time of Day Pricing of Electrical Energy: Does It Promote the National Interest?* Upton, N.Y., October.

Burright, B., J. Enns, T. F. Kirkwood, and S. Wildhorn (1974), "How to Save Gasoline: Public Policy Alternatives for the Automobile," Rand Corporation (R-1560), Santa Monica, Calif.

California (1979), *Solar Tax Credit Regulations* (as amended July 11).

California, Department of Transportation (1979), *California Ridesharing Program*, Second Interim Report (March).

California Energy Commission (CEC) (undated, a), "California and Federal Solar Tax Credit Laws. Summary and Comparison" (information sheet).

California Energy Commission (CEC) (undated, b), "California Solar Tax Credit," brochure.

California Energy Commission (CEC) (1976), *Reducing Energy Requirements for Residential Water Heating*, December.

California Energy Commission (CEC) (1977a), *California Energy Trends and Choices*, Vol. 3 (biennial report of the CEC).

California Energy Commission (CEC) (1977b), *The National Energy Plan: Implications for California*, May 18.

California Energy Commission (CEC) (1978), *Joint Investigation by the California Energy Commission and the California Public Utilities Commission into the Availability and Potential Use of Solar Energy in California*, Decision, October 12.

California Energy Commission (CEC) (1978a), *Guidelines and Criteria for the California Solar Energy Tax Credit*, April.

California Energy Commission (CEC) (1978b), *The Warren-Alquist Act*, December.

California Energy Commission (CEC) (1978c), *California Residential Energy Consumption Survey*, Final Report by Selection Consulting Center, September.

California Energy Commission (CEC) (1978d), *Guidelines for Certification of Solar Energy Equipment*, June 15.

California Energy Commission (CEC) (1979), *California Marginal Cost Pricing Project, A First Year Report to the Department of Energy*, September.

California Energy Commission (CEC) (1979a), *Energy Choices for California. Looking Ahead*, March.

California Energy Commission (CEC) (1979b), *Interpreting the Results of the Atlas Report*, "Energy Savings Resulting from Incremental Insulation and Incremental Solar System Capacity," Relative to AB 1558 Tax Credit Guidelines, December 14.

California Energy Commission (CEC) (1979c), *California Load Management Research 1978*, May.

California Energy Commission (CEC) (1979d), Consultant Report (Dr. Duane Chapman), *Taxation and Solar Energy*, June.

California Energy Commission (CEC) (1979e), *Exploring New Energy Choices for California*, The 1979/80 R&D Report to the Legislature, June.

California Energy Commission (CEC) (1979f), *Exploring New Energy Choices for California*, The 1979/80 R&D Report to the Legislature, Vol. 1, June.

California Energy Commission (CEC) (1979g), *California Energy Demand 1978–2000, A Preliminary Assessment*, August.

California Energy Commission (CEC) (1979h), *Toward An Alternative Energy Path for California: A Preliminary Action Agenda*, August.

California Energy Commission (CEC), Conservation Division, Transportation Element (1979i), *1979–1980, Budget Workplan*, unpublished draft, August 2.

California Energy Commission (CEC) (1979j), *California's Solar Energy Tax Credit: An Analysis of Tax Returns for 1977*, Staff Report, September.

California Energy Commission (CEC) (1979k), *New Initiatives for Energy Conservation*, Part 1, August (staff draft).

California Energy Commission (CEC) (1981), *Energy Tomorrow, Challenges and Opportunities for California*, Biennial Report.

California, Secretary of State (1979), *Administrative Regulations* filed June 9, Load Management Standards.

California Transportation Commission (CTC) (1979), *Financing Transportation in California. Vol. 1: Recommendation*, April 27.

California Utilities Solar Programs (undated), unpublished information compiled from hearing testimony before the PUC.

Camm, Frank A. (1978), *Average Cost Pricing of Natural Gas: A Problem and Three Policy Options*, Rand Corporation (R-2282), Santa Monica, Calif., July.

Canadian Electrical Association (1978), *Marginal Costing and Pricing of Electrical Energy, A State-of-the-Art Conference*, May.

Carpenter, W. W. (1978), "Alternatives to Marginal Cost Pricing," in Canadian Electrical Association, *Marginal Costing and Pricing of Electrical Energy, A State-of-the-Art Conference*, May, pp. 21–37.

Caves, Douglas W., and Laurits R. Christensen (1978a), "Time-of-Use Pricing for Residential Electricity Customers: The Wisconsin Experiment," Prepared for presentation at *Marginal Costing and Pricing of Electrical Energy, A State-of-the-Art Conference* sponsored by the Canadian Electrical Association, Montreal, Quebec, Canada, May 1–4.

Caves, Douglas W., and Laurits R. Christensen (1978b), "Econometric Analysis of the Wisconsin Residential Time-of-Use Electricity Pricing Experiment," Social Systems Research Institute, University of Wisconsin, Madison, September.

Caves, Douglas W., and Laurits R. Christensen, (1979), "Residential Substitution of Off-Peak for Peak Electricity Usage Under Time-of-Use Pricing: An Analysis of 1976 and 1977 Summer Data from the Wisconsin Experiment," April 9.

Caves, Douglas W., and L. R. Christensen (1980), "Residential Substitution of Off-Peak for Peak Electricity Usage Under Time-of-Use Pricing," *The Energy Journal*, 1(2), 85–142.

Chamberlain, C. (1973), "Models of Gasoline Demand," U.S. Department of Transportation, Transportation Systems Center, Cambridge, Mass.

Chern, W. S. (1978), "Aggregate Demand for Energy in the United States," in G. S. Maddala and others (eds.), *Econometric Studies in Energy Demand and Supply*, Praeger, New York, pp. 5–43.

Christensen, Laurits R., and W. H. Green (1978), "Using Statistical Cost Analysis to Estimate Marginal Costs for Electric Utilities," in Canadian Electrical Association, *Marginal Costing and Pricing of Electrical Energy, A State-of-the-Art Conference*, May, pp. 107–113.

Cicchetti, Charles J., and Mo Reinbergs (1979), "Electricity and Natural Gas Rate Issues," *Annual Review of Energy*, 4, 231–258.

Cicchetti, Charles J., and V. Kerry Smith (1975), "Alternative Price Measures and the Residential Demand for Electricity: A Specification Analysis," *Regional Science and Urban Economics*, 5(4), 503–516.

Cicchetti, Charles J., William J. Gillen, and Paul Smolensky (1977), *The Marginal Cost and Pricing of Electricity: An Applied Approach*, Ballinger, Cambridge, Mass.

Clark, Norm (1979), "Volume of Alternative Energy Home Loans," Oregon Department of Veterans' Affairs, Interoffice Memo, November 15.

Connecticut Public Utilities Control Authority (PUCA) (1977), *Connecticut Peak Load Pricing Test, Final Report*, May.

Council of State Government (1976), *State Energy Management: The California Energy Resources Conservation and Development Commission*, Lexington, KY., May.

Denny, M. G. S., and C. Pinto (1975), *The Demand for Energy in Canadian Manufacturing: 1949–1970*, mimeographed paper, University of Toronto.

Dewees, D. N., R. M. Hyndman, and L. Waverman (1974), "The Demand for Gasoline in Canada 1956–1972," University of Toronto, Institute for the Quantitative Analysis of Social and Economic Policy, Working Paper No. 7406, June.

Doctor, R. D., et al. (1972), *California's Electricity Quandary: III, Slowing the Growth Rate*, Rand Corporation, Santa Monica, Calif.

Dole, Stephen H. (1975), *Energy Use and Conservation in the Residential Sector: A Regional Analysis*, Rand Corporation, Santa Monica, Calif.

EBASCO Services Incorporated (1977a), *Costing for Peak Load Pricing: Topic 4*, Prepared for EPRI Rate Design Study, May 4.

EBASCO Services Incorporated (1977b), *The Development of Various Pricing Approaches: Topic 1.3*, Prepared for EPRI Rate Design Study, March 1.

Echolm, Art (1977), "New Rate Structures in Texas," *Federal Reserve Board of Dallas Review*, November.

Edmonson, Nathan (1975), "Real Price and the Consumption of Mineral Energy in the United States, 1901–1968," *Journal of Industrial Economics*, **23**(3), 161–174.

Efford, Ian E. (1978), "Marginal Cost Pricing from an Energy Conservation Perspective," in Canadian Electrical Association, *Marginal Costing and Pricing of Electrical Energy*, A State-of-the-Art Conference, May, pp. 59–65.

Electricity Consumers Resource Council (1978), "Public Utility Regulatory Policies Act," November.

Elliott, Ward (1975), "The Los Angeles Affliction: Suggestions for a Cure," *The Public Interest*, Winter pp. 119–128.

Energy, Mines and Resources Canada (EM&R) (1976), *An Energy Strategy for Canada: Policies for Self-Reliance*, Ottawa.

Energy, Mines and Resources Canada (EM&R) (1978a), *Energy Futures for Canadians*, Ottawa.

Energy, Mines and Resources Canada (EM&R) (1978b), *Solar Heating and Employment in Canada*, Report ER 79-1, September.

Energy, Mines and Resources Canada (EM&R) (1980), *Electric Power in Canada, 1979*.

Epstein, P., and D. Barrett, (1977), *Federal Incentives for Solar Homes: An Assessment of Program Options*, U.S. Department of Housing and Urban Development, Washington, D.C.

Erickson, E. W., R. M. Spann, and R. Ciliano (1973), "Fossil Fuel Energy Demand Analysis," Report to the Office of Science and Technology, Decision Sciences Corp., Washington, D.C., March 20.

Ernst & Ernst (1977), *The Long-Run Incremental Cost of Electric Utilities (Final Draft Report)*. Prepared for National Science Foundation, Washington, D.C., October.

The Financial Post (1980), "Electric Power", Special Report, February 9.

Foster, Harold D., and Derrick Sewell (1977), *Solar Heating in Canada: Problems and Prospects*, Fisheries and Environment Canada.

Fuss, M. A. (1975), *The Demand for Energy in Canadian Manufacturing: An Example of the Estimation of Production Structure with Many Inputs*, Institute for Policy Analysis, University of Toronto.

Fuss, Melvyn A. (1977), *The Derived Demand for Energy in the Presence of Supply Constraints*, Institute for Policy Analysis, University of Toronto, Working Paper No. 7714, August.

Fuss, M., and L. Waverman. (1975a), *The Demand for Energy in Canada,* report to Energy, Mines and Resources Canada, February.

Fuss, M., and L. Waverman (1975b), *The Demand for Energy in Canada,* Institute for Policy Analysis, University of Toronto, February, 3 vols., Report Series No. 7.

Fuss, M., R. Hyndman, and L. Waverman (1977), "Residential, Commercial and Industrial Demand for Energy in Canada: Projections to 1985 with Three Alternative Models," in William D. Nordhaus (ed.), *International Studies of the Demand for Energy,* North-Holland, Amsterdam, pp. 151–179.

Gallagher, Thomas J., Jr. (1975–1976), "An Automobile Fuel Consumption Efficiency Tax: A Sumptuary Response to the 'Efficiency Crisis,' " *Rutgers Camden Law Journal,* **7.**

GMA Research Corp. (1979), *Oregon Residential Energy Conservation Survey; Executive Summary.* March.

Gorbet, F. W. (1975), "Energy Demand Projections for Canada: An Integrated Approach," Paper presented at the American Institute of Mining Engineers Meetings, New York, February 16–20 (available from EM&R, Canada).

Gordian Associates, Inc. (undated, a), "Brief Description of the Gordian Associates Electric Utility System Model," New York.

Gordian Associates, Inc. (undated, b), "Description of the Gordian Computer Model for Electric Utility Expansion Planning," New York.

Gordian Associates, Inc. (1976), *UCAN Technical Implementation Manual,* prepared for U.S. Federal Energy Administration, August.

Gordian Associates, Inc. (1978), "Summary Description of the Detailed RP950 Electricity Supply Model," Mimeo prepared for EPRI, May, New York.

Grandjean, Burke D., and Patricia A. Taylor (1976), "Public Policy and Renters' Electric Bills," *Social Science Quarterly,* **57**(2), 437–444.

Gray, Larry (1979), "Evaluation of Oregon's Energy Supplier Weatherization Program," State of Oregon, Interoffice Memo, July 16.

Halvorsen, Robert (1975), "Residential Demand for Electric Energy," *Review of Economics and Statistics,* **57**(1), 12–18.

Hartman, Raymond S. (1979), "Frontiers in Energy Demand Modeling," in J. Hollander, M. Simmons, and D. Wood (eds.), *Annual Review of Energy,* **4,** Annual Reviews Inc., Palo Alto, Calif.

Heberlein, Thomas A., and Keith Warriner (1978), "Appliance Saturation, Housing and Demographic Characteristics of Residential Customers," Department of Rural Sociology, University of Wisconsin, Madison, March 10.

Helliwell, J. F. (1978), "Some Emerging Economic Issues in Utility Regulation and Rate-Making," *Logistics and Transportation Review,* **15.**

Helliwell, John F. (1979), "Canadian Energy Policy," *Annual Review of Energy,* **4,** Annual Reviews Inc., Palo Alto, Calif., 175–229.

Hemphill, Marion (1979), Personal Communication, August.

Hendricks, Wallace, Roger Koenker, and Robert Pedlasek (1977), "Consumption Patterns for Electricity," *Journal of Econometrics,* **5**(2), 135–153.

Hirst, Eric, and Janet Carney (1978), "Effects of Federal Residential Energy Conservation Programs," *Science,* **199,** February 24.

Houthakker, H. S., and M. Kennedy (1973), "The World Demand for Petroleum Model," Data Resources, Inc., Lexington, Mass.

Houthakker, H. S., and Lester D. Taylor (1970), *Consumer Demand in the United States: Analyses and Projections,* Harvard University Press, Cambridge, Mass.

Houthakker, H. S., and P. Verleger (1973), "The Demand for Gasoline: A Mixed Cross-Sectional and Time Series Analysis," Data Resources, Inc., Lexington, Mass.

Houthakker, H. S., P. K. Verleger, Jr., and D. P. Sheehan (1974), "Dynamic Demand Analyses for Gasoline and Residential Electricity," *American Journal of Agricultural Economics*, **56**(2), 412–418.

Hunt, Marshall, and David Bainbridge (1978), "The Davis Experience," *Solar Age*, (May) 20–23.

Hyatt, Sherry V. (1977), "Thermal Efficiency and Taxes: The Residential Energy Conservation Tax Credit," *Harvard Journal on Legislation*, **14**(2).

Hyndman, R., and G. F. Mathewson (1975), *A Report to Ontario Hydro on the Residential Demand for Energy in Canada*, October 15.

Joiner, Brian L., and Ian Ford (1978), "Design of the Wisconsin Time-of-Day Electrical Pricing Project," Department of Statistics, University of Wisconsin, Madison, April.

Joskow, P. L., and M. L. Baughman (1975), "The Future of the U.S. Nuclear Energy Industry," Working Paper No. 155, Department of Economics, Massachusetts Institute of Technology, Cambridge, Mass., April.

Khazzoom, J. Daniel (1973), "An Econometric Model of the Demand for Energy in Canada, Part One: The Industrial Demand for Gas," *The Canadian Journal of Statistics*, **1**(1), 69–107.

Khazzoom, J. Daniel (1974), "An Econometric Model of the Demand for Energy in Canada," *Energy: Demand, Conservation and Institutional Problems*, in Michael S. Macrakis (ed.), Massachusetts Institute of Technology, Cambridge, Mass., pp. 360–374.

Khazzoom, J. Daniel (1977), "An Application of the Concepts of Free and Captive Demand to the Estimating and Simulating of Energy Demand in Canada," in William D. Nordhaus (ed.), *International Studies of the Demand for Energy*, North-Holland, Amsterdam, pp. 115–136.

Kiphut, Alan D (1978), "Oregon's Solar Tax Credit Program—The First 100 Installations," Solar 78 Northwest, Conference Proceedings, Portland, July 14–16.

Knelman, Fred H. (1975), *Energy Conservation*, Science Council of Canada, Background Study No. 3, Ottawa, Canada.

Kraft, John, and Mark Rodekohr (1978), "Regional Demand for Gasoline: A Temporal Cross-Section Specification," *Journal of Regional Science*, **18**(1), 45–56.

Lacy, A. W., and D. R. Street (1975), "A Single Firm Analysis of the Residential Demand for Electricity," Department of Economics, Auburn University, Auburn, Alabama.

Lady, G. (1974), "National Petroleum Product Supply and Demand," Federal Energy Administration, Technical Report 74-5, Washington, D.C.

Lehman, Richard, and Henry Warren (1978), "Residential Natural Gas Consumption: Evidence That Conservation Efforts to Date Have Failed," *Science*, **199**, February 24.

Little, Arthur D., Inc. (1975), *Efficacy of Incentives to Optimize Performance and Efficiency of Air Conditioning: Report on Phase 1 Study*, National Science Foundation, Washington, D.C.

Lyman, R. Ashley (1978), "Price Elasticities in the Electric Power Industry," *Energy Systems and Policy*, **2**(4), 381–406.

Malko, J. Robert (1978), "Implementing Time-of-Use Pricing," presented at the Engineering Economy for Public Utilities, Seventeenth Annual Program, Stanford University, Palo Alto, Calif., July.

Malko, J. Robert, and T. B. Nicolai (1983), "Implementing Residential Time-of-Day Pricing of Electricity in Wisconsin: Some Current Activities and Issues," 9th Annual Rate Symposium, Problems of Regulated Industries, Institute for Study of Regulation, University of Missouri-Columbia, February.

Malko, J. R., Malcolm A. Lindsay, and Carol T. Everett (1977), "Towards Implementation of Peak-Load Pricing of Electricity: A Challenge for Applied Economics," *The Journal of Energy and Development*, (Autumn).

Malko, J. Robert, Dennis J. Ray, and Nancy L. Hassig (1981), "Time-of-Day Pricing of Electricity Activities in Some Midwestern States," in Peter N. Nemetz (ed.), *Energy Crisis: Policy Response*, Institute for Research on Public Policy, Montreal, pp. 143–170.

Manning, Willard G., Bridger M. Mitchell and Jan P. Acton (1979), "Design of the Los Angeles Peak-Load Pricing Experiment for Electricity," *Journal of Econometrics*, Vol. 2, No. 1, September, pp. 131–194.

Mathewson, G. F., and Associates (1976), "Residential Demand for Electric Energy and Natural Gas: A General Model Estimated for Canada with Forecasts," *Ontario Hydro, Electricity Costing and Pricing Study*, 4 (October), 1–196.

McGillivray, R. G. (1974), "Gasoline Use by Automobiles," Urban Institute, Working Paper 1216-2, Washington, D.C.

McNeal, Hildebrand, and Associates Ltd. (1979), *Market Study of Insulation Products in the Pacific Northwest*, Vancouver, B.C., April.

Minan, John H., and William H. Lawrence (1978), "State Tax Incentives to Promote the Use of Solar Energy," *Texas Law Review*, 56(791), 835–859.

Mitchell, Bridger M., and J. P. Acton (1977), *Peak-Load Pricing in Selected European Electric Utilities*, Rand Corporation (R-2031-DWP), Santa Monica, Calif., July.

Mitchell, Bridger M., W. G. Manning, and J. P. Acton (1977), *Electricity Pricing and Load Management: Foreign Experience and California Opportunities*, Rand Corporation (R-2106-CERCDC), Santa Monica, Calif., March.

Mount, T. D., and L. D. Chapman (1976), "Effects of Increasing the Use of Electricity on Environmental Quality in the U.S.: A Model of Power Generation and the Policy Issues Raised by Its Application," in W. D. Nordhaus (ed.), *Proceedings of the Workshop on Energy Demand*, International Institute for Applied Systems Analysis, Laxenburg, Austria.

Mount, T. D. and others (1974), "Electricity Demand in the United States: An Econometric Analysis," in Michael S. Macrakis (ed.), *Energy: Demand, Conservation, and Institutional Problems*, Massachusetts Institute of Technology, Cambridge, Mass., pp. 318–329.

National Association of Regulatory Utility Commissioners (NARUC) (1977), *Electric Utility Rate Design Study, Rate Design and Load Control*, Washington, D.C., November.

National Economic Research Associates, Inc. (NERA) (1975), "Lifeline Rates and Energy Stamps," New York, pamphlet.

National Economic Research Associates (NERA) (1976), *Analysis of Price Elasticity of Demand for Electricity in Ontario Hydro's Service Area*, report to Ontario Hydro, March, (reproduced in Ontario Hydro, 1976d).

National Economic Research Associates, Inc. (NERA) (1977a), *A Framework for Marginal Cost-Based Time Differentiated Pricing in the U.S.*, New York, Topic 1.3, February 21.

National Economic Research Associates (NERA) (1977b), *How to Quantify Marginal Costs: Topic 4*. Prepared for EPRI Rate Design Study, New York, March 10.

National Regulatory Research Institute (1978), *The Missouri Rate Reform Initiative*, Ohio State University: Columbus, Ohio, December.

National Science Foundation (1975), *Tax Policy and Energy Conservation*, Washington, D.C., March.

Nelson, Jon P. (1975), "The Demand for Space Heating Energy," *Review of Economics and Statistics*, 57(4), 508–512.

Nelson, Samuel H. (Argonne National Laboratory) (1976), *Strategies for Commercializing Customer Thermal-Energy Storage*, Lemont, Illinois, December.

Nemetz, Peter N. (1979), *Economic Incentives for Energy Conservation at the Consumer*

Level—An Overview and Preliminary Synthesis, Report for Department of Consumer and Corporate Affairs, Ottawa.

Northwest Power Planning Council (1983a), *Regional Conservation and Electric Power Plan* (draft), January 26.

Northwest Power Planning Council (1983b), *Northwest Conservation and Electric Power Plan*, April 27.

Ontario Energy Board (OEB) (1979), *Report to the Minister of Energy on Principles of Electricity Costing and Pricing for Ontario Hydro*, H.R. 5, December 20.

Ontario Hydro (1976a), Electricity Costing and Pricing Study (ECAPS), *Study Overview and Principal Policy Recommendations, Vol. I, October.*

Ontario Hydro (1976b), Electricity Costing and Pricing Study (ECAPS), *Determining the Annual Cost of Power and Report on Allocation of Costs*, Vol. II, October.

Ontario Hydro (1976c), Electricity Costing and Pricing Study (ECAPS), *Report on Inflation Accounting*, Vol. III, October.

Ontario Hydro (1976d), Electricity Costing and Pricing Study (ECAPS), *The Demand for Electricity*, Vol. IV, October.

Ontario Hydro (1976e), Electricity Costing and Pricing Study (ECAPS), *Theoretical Foundations of Marginal Cost Pricing*, Vol. V, October.

Ontario Hydro (1976f), Electricity Costing and Pricing Study (ECAPS), *Alternative Objectives for Pricing*, Vol. VI, October.

Ontario Hydro (1976g), Electricity Costing and Pricing Study (ECAPS), *Costing Methodology for Determining Marginal Costs*, Vol. VII, October.

Ontario Hydro (1976h), Electricity Costing and Pricing Study (ECAPS), *Detailed Rate Structure Design Proposals*, Vol. VIII, October.

Ontario Hydro (1976i), Electricity Costing and Pricing Study (ECAPS), *Interruptible Power Study*, Vol. IX, October.

Ontario Hydro (1976j), Electricity Costing and Pricing Study (ECAPS), *Impact Study*, Vol. X, October.

Oregon Department of Energy (ODOE) (no date, a), *Saving Energy, Saving Money; It Makes Sense!* Salem.

Oregon Department of Energy (ODOE) (no date, b), *Weatherization Incentives for Oregonians*, Salem.

Oregon Department of Energy (ODOE) (1977a), *Something New! 1977 Weatherization Incentives for Oregon Homeowners*, Salem, August.

Oregon Department of Energy (ODOE) (1977b), *Energy Conservation Legislation Final Summaries*, September 27, unpublished.

Oregon Department of Energy (ODOE) (1978a), *Amended Oregon State Energy Conservation Plan*, Salem, April.

Oregon Department of Energy (ODOE) (1978b), "Tax Credit Eligibility Criteria for Residential Alternative Energy Devices: Solar, Geothermal, and Wind," Chapter 330, unpublished.

Oregon Department of Energy (ODOE) (1978c), *Oregon's Energy Future, Second Annual Report*, January 1.

Oregon Department of Energy (ODOE) (1979a), *Amended Oregon State Energy Conservation Plan*, Salem, May.

Oregon Department of Energy (ODOE) (1979b), *Amended Oregon State Energy Conservation Plan*, Salem, August (draft version).

Oregon Department of Energy (ODOE) (1979c), "Energy Measures Approved; the 1979 Oregon Legislature," Salem, unpublished.

Oregon Department of Energy (ODOE) (1979d), *Oregon's Energy Future*, Third Annual Report, January 1.

Oregon Department of Energy (ODOE) (1982a), "Energy Savings from State Programs" unpublished memo, July 29.

Oregon Department of Energy (ODOE) (1982b), *Energy Management in State Building*, Report for 1981, October.

Oregon Department of Energy (ODOE) (1983a), "Residential Tax Credit Program," unpublished memo, January 14.

Oregon Department of Energy (ODOE) (1983b), "Energy Legislation Summaries for 1981," Salem, January 17.

Oregon Department of Energy (ODOE) (1983c), "Oil Heat Weatherization Program," unpublished memo, January 18.

Oregon Department of Eneregy (ODOE) (1983d), *Seventh Annual Report*, January.

Oregon Department of Revenue (1978), "Weatherization Refunds for Seniors," Information Circular, Salem, October.

Oregon Department of Revenue (1979a), "Tax Credit Weatherization," Information Circular, Salem, April.

Oregon Department of Revenue (1979b), "Tax Credit for Weatherization," Information Circular, Salem, October.

Oregon Department of Revenue (1979c), "Income Tax Credit for Alternative Energy Devices," Information Circular, Salem, December.

Oregon Department of Revenue (1980), Personal Communication, January 7.

Oregon Department of Veterans' Affairs (1981), "Loan Fact Sheet," January 1.

Oregon Legislative Assembly (1977a), Enrolled House Bill 2156, Chapter 383, Regular Session.

Oregon Legislative Assembly (1977b), Enrolled House Bill 2157, Chapter 889, Regular Session.

Oregon Legislative Assembly (1977c), Enrolled House Bill 2701, Chapter 811, Regular Session.

Oregon Legislative Assembly (1977d), Enrolled House Bill 3265, Chapter 887, Regular Session.

Oregon Legislative Assembly (1977e), Enrolled Senate Bill 4, Chapter 716, Regular Session.

Oregon Legislative Assembly (1977f), Enrolled Senate Bill 339, Chapter 196, Regular Session.

Oregon Legislative Assembly (1977g), Enrolled Senate Bill 477, Chapter 315, Regular Session.

Oregon Legislative Assembly (1981a), Senate Bill 11.

Oregon Legislative Assembly (1981b), House Bill 2246, Chapter 778.

Oregon Legislative Assembly (1981c), House Bill 2247, Chapter 894.

Oregon Legislative Assembly (1981d), House Bill 2248, Chapter 565.

Oregon, Public Utility Commissioner (1978), *1977 Statistics of Oregon Public Utilities*, Salem, October.

Oregon, Public Utility Commissioner (1981a), *Oregon Utility Statistics*, Salem.

Oregon, Public Utility Commissioner (1981b), *Residential Weatherization Program*, Salem, December.

Oregon, Public Utility Commissioner (1982), *1981 Statistics of Oregon Public Utilities*, Salem.

Oregon State Executive Department (1973), *Oregon's Energy Perspective*. Salem, May.

Osawa, E. (1978), "Electricity Tariff System in Japan," *Marginal Costing and Pricing of Electrical Energy, A State-of-the-Art Conference*, Canadian Electrical Association, May, pp. 92–106.

Osler, Sanford (1977), "An Application of Maginal Cost Pricing Principles to B.C. Hydro," UBC Department of Economics, Resource Paper No. 12, July.

Pacific Northwest Electric Power Planning and Conservation Act, 1980 (1981), Bonneville Power Administration, U.S. Department of Energy.

Pacific Power & Light Company (PP&L) (1978a), "Background Information; Pacific Power's Weatherization Loan Program," Columbia Basin Division Office, Yakima, Washington.

Pacific Power & Light Company (PP&L) (1978b), "Proposed Residential Energy Efficiency Rider," Testimonies before the Public Utility Commissioner of the State of Oregon, April.

Pacific Power & Light Company (PP&L) (1979a), *Pocket Facts*, Portland, Oregon.

Pacific Power & Light Company (PP&L) (1979b), *Residential Electric Bills*, April.

Pacific Power & Light Company (PP&L) (1982), *Residential Rate History 1977–1982*.

Peck, A. E. and O. C. Doering III (1976), "Voluntarism and Price Response: Consumer Reaction to the Energy Shortage," *Bell Journal of Economics and Management Science*, 7(1), 287–292.

Pindyck, R. S. (1976), *International Comparisons of the Residential Demand for Energy: A Preliminary Analysis (World Oil Project)*. MIT Energy Lab for National Science Foundation, Cambridge, Mass., September.

The City of Portland (undated), *Proposed Energy Policy for Portland*, Discussion Draft.

The City of Portland (1979), Ordinance 148251. *An Ordinance Adopting an Energy Conservation Policy for Portland*, August 15.

Portland Bureau of Planning (1977a), *Energy Conservation Choices for the City of Portland*. June, 11 vols.

Portland Bureau of Planning (1977b), *Energy Conservation Choices for the City of Portland, Vol. 3, Summary of Conservation Choices*, June.

Portland Bureau of Planning (1977c), *Energy Conservation Choices for the City of Portland, Vol. 3A, Residential Conservation Choices*, June.

Portland Bureau of Planning (1977d), *Energy Conservation Choices for the City of Portland, Vol. 3B, Transportation and Land Use Conservation Choices*, June.

Portland Bureau of Planning (1977e), *Energy Conservation Choices for the City of Portland, Vol. 6, Project Overview*, June.

Protti, Gerard J. (1978), "The Potential Impact of Marginal Cost Pricing on Canadian Electric Utility Expenditures," *Marginal Costing and Pricing of Electrical Energy, A State-of-the-Art Conference*, Canadian Electrical Association, May, pp. 268–275.

Ramsey, J., and others (1975), "An Analysis of the Private and Commercial Demand for Gasoline," *Review of Economics and Statistics*, 57(4), 502–507.

Rand Corporation (1972), *California's Electricity Quandary*, Santa Monica, Calif., September.

Randall, A., B. C. Ives, and J. T. Ryan (1974), "The Demand for Electricity and Natural Gas in the Southwest," New Mexico State University, Las Cruces.

Ray, Dennis, and Rodney Stevenson (1980), "Measuring the Potential Impacts from Lifeline Pricing of Electricity and Natural Gas Services," in Michael A. Crew (ed.), *Issues in Public Utility Pricing and Regulation*, Lexington Books, Lexington, Mass.

Ray, Dennis J., J. Stanley Black, and J. Robert Malko (1978), "Developing and Implementing a Peak-Load Pricing Experiment for Residential Electricity Customers: A Wisconsin Experience," prepared for the Midwest Economics Association Annual Meeting, Chicago, April 6.

Ray, Dennis, J. Kaul, and J. R. Malko (1980), "Estimating Usage Response of Wisconsin Industrial Customers to Time-of-Day Electricity Rates: A Preliminary Analysis," Midwest Economics Association Annual Meeting, March.

Ruffell, R. J. (1978), "Measurement of Own-Price Effects on the Household Demand for Electricity," *Applied Economics*, 10(1), 21–30.

Schiflett, Mary, and John V. Zuckerman (1978), "Solar Heating and Cooling: State and Municipal Legal Impediments and Incentives," *Natural Resources Journal*, 18(2).

Seidel, Marquis, R., et al. (1973), *Energy Conservation Strategies*, U.S. Environment Protection Agency.

Solar Energy Research Institute (SERI) (1979a), *The Implementation of State Solar Incentives: A Preliminary Assessment*, Golden, Colorado, January.

Solar Energy Research Institute (1979b), *Turning Laws into Incentives: The Implementation of State Solar Energy Initiatives*, Colorado, February.

Solar Use Now for Resources and Employment (SUNRAE) (undated), *Five Year Survey of California Solar Legislation (1973–1978)*, Sacramento, Calif.

Spurlock, C. W. (1978), "Forecasting Regional Demand for Heating Fuel," *Growth and Change*, 9(2), 29–34.

Statistics Canada (1978a), *Canada Year Book 1978–79*. Minister of Supply and Services, Ottawa, Canada.

Statistics Canada (1978b), *Road Motor Vehicles, Fuel Sales, 1977*, Catalogue 53–218, Ottawa, Canada, September.

Sullivan, Timothy, J. (1979), *The Los Angeles Senior Citizen Lifeline Electricity Rate*, Rand Corporation, Santa Monica, Calif., January.

Sweeney, J, (1975), "A Vintage Capital Stock Model of Gasoline Demand," U.S. Federal Energy Administration, Washington, D.C.

Taylor, Lester D. (1975), "The Demand for Electricity: A Survey," *Bell Journal of Economics and Management Science*, 6(1), 74–110.

Taylor, Lester D. (1977), "The Demand for Energy: A Survey of Price and Income Elasticities," in William D. Nordhaus (ed.), *International Studies of the Demand for Energy*, North-Holland, Amsterdam, pp. 3–43.

Taylor, Lester, et al. (1977) *The Welfare Effects of Fuel Economy Policies*, Data Resources Inc., Lexington, Mass.

Taylor, L. D., G. R. Blattenberger, and P. K. Verleger (1976), "The Residential Demand for Energy," Report to the Electric Power Research Institute, Department of Economics, University of Arizona, Tucson.

Turvey, Ralph (1978), "The Case for Marginal Cost Pricing of Electricity," *Marginal Costing and Pricing of Electrical Energy, A State-of-the-Art Conference*, Canadian Electrical Association, May, pp. 1–7.

Tyrrell, T. J. (1974), "Projections of Electricity Demand," in Michael S. Macrakis (ed.), *Energy: Demand, Conservation and Institutional Problems*, Massachusetts Institute of Technology, Cambridge, Mass., 1974, pp. 342–359.

U.S. Congress, Congressional Budget Office (1979), *The Decontrol of Domestic Oil Prices: An Overview*, May.

U.S. Community Services Administration (1977a), *The Weatherization Program: A Policy Perspective*, April.

U.S. Community Services Administration (1977b), *The Feasibility of an Energy-Related Loan Program for Low-Income Homeowners*, October.

U.S. Department of Commerce, (NTIS) (1978), *Electric Power Load Management, A Bibliography with Abstracts, 1964–July 1978.*

U.S. Department of Energy (DOE) (1977), *Energy Northwest.* October.

U.S. Department of Energy (DOE) (1978a), *Analysis of the Effects of the Time-of-Use Electricity Rate in the Ohio Electric Utility Demonstration Project, Draft Final Report*, August.

U.S. Department of Energy (DOE) (1978b), *Time-of-Use Electricity Prices, Arizona*, December.

U.S. Department of Energy (DOE) (1978c), *Electric Utility Rate Design and Energy Management Initiatives*, 1977 Annual Report, DOE/ERA-0011, January.

U.S. Department of Energy (DOE) (1978d), "Information Kit; The National Energy Act," Office of Public Affairs, November.

U.S. Department of Energy (DOE) (1978e), *Projected 1985 Automobile Gasoline Savings from the House-Passed Gas Guzzler Tax and Conference Minimum Mileage Standards*, DOE/EIA-0102/3, June.

U.S. Department of Energy (DOE) (1978f), *An Evaluation of Energy Related Tax and Tax Credit Programs*, AM/IA/78-12, July 17.

U.S. Department of Energy (DOE) (1979a) *Institutional Analysis of Alternative Electric Rate Designs and Related Regulatory Issues in Support of DOE Utility Conservation Programs and Policy, Vol. IV, Institutional Analysis*, May.

U.S. Department of Energy (DOE) (1979b), *Technical, Institutional and Economic Analysis of Alternative Electric Rate Designs and Related Regulatory Issues in Support of DOE Utility Conservation Programs and Policy, Vol. I, Domestic Rate Survey*, May.

U.S. Department of Energy (DOE) (1979c), *Technical, Institutional and Economic Analysis of Alternative Electric Rate Designs and Related Regulatory Issues in Support of DOE Utility Conservation Programs and Policy, Vol. II, Foreign Rate Survey*, May.

U.S. Department of Energy (DOE) (1979d), *Technical, Institutional and Economic Analysis of Alternative Electric Rate Designs and Related Regulatory Issues in Support of DOE Utility Conservation Programs and Policy, Vol. III, Economic Analysis*, May.

U.S. Department of Energy (DOE) (1979e), *Time-of-Use Electricity Price Effects—Ohio.*

U.S. Department of Energy (DOE) (1979?) *Arkansas Demand Management Demonstration Study.*

U.S. Department of Transportation, National Highway Traffic Safety Administration (1980), *Automobile Fuel Economy Program*, Fourth Annual Report to Congress, January.

U.S. Environmental Protection Agency (EPA) (1973), *Energy Conservation Strategies*, EPA-R5-73-021, July.

U.S. Environmental Protection Agency (EPA) (1978), *Vanpooling: An Overview*, Denver, Colo., March.

U.S. Energy Research and Development Administration (ERDA) (1976), *Life Cycle Costing Emphasizing Energy Conservation*, September.

U.S. Federal Energy Administration (FEA) (1976a), *Attitudes, Knowledge and Behavior of American Consumers Regarding Energy Conservation with Some Implications for Governmental Actions.*

U.S. Federal Energy Administration (FEA) (1976b), *Impact of Load Management Strategies Upon Electric Utility Costs and Fuel Consumption*, August.

U.S. Federal Energy Administration (FEA) (1976c), *Carpool Incentives: Evaluation of Operational Experience*, March.

U.S. Federal Energy Administration (FEA) (1976d), *1976 National Energy Outlook*, Washington, D.C.

U.S. Federal Energy Administration (FEA) (1977a), *Electric Utility Rate Design Proposals*, February.

U.S. Federal Energy Administration (FEA) (1977b), *Investigations into the Effects of Rate Structure on Customer Electric Usage Patterns, Final Report, State of Vermont, Public Service Board*, March.

U.S. Federal Energy Administration (FEA) (1977c), *Lender Impacts upon Energy Conservation in Buildings*, FEA/D-77/125, February.

U.S. Federal Energy Administration (FEA) (1977d), *State Energy Conservation Program Sourcebook. Vol. 2, State Energy Conservation Plan Handbook*, January.

U.S. Federal Energy Administration (FEA) (1977e), *Mode Shift Strategies to Effect Energy Savings in Intercity Transportation*, April.

U.S. 94th Congress (1975), S.622. Energy Policy and Conservation Act. Public Law 94–163, December 22.

U.S. 94th Congress (1976), H.R. 12169. Energy Conservation and Production Act. Public Law 94–385, August 14.

U.S. Senate (1977), Energy Production and Conservation Tax Incentive Act (Report of the Committee on Finance on H.R. 5263).

Uri, N. D. (1975), "A Dynamic Demand Analysis for Electrical Energy by Class of Consumer," Working Paper 34, Bureau of Labor Statistics, January.

Verleger, P. (1973), *A Study of the Quarterly Demand for Gasoline and Impacts of Alternative Gasoline Taxes*, Data Resources Inc., Lexington, Mass.

Verleger, P. K., and D. P. Sheehan (1974), "The Demand for Distillate Fuel Oil and Residual Fuel Oil: A Cross-Section Time-Series Study," Report prepared for the U.S. Council on Environmental Quality, Data Resources, Inc., Lexington, Mass.

Vickrey, William (1978), "Efficient Pricing Under Regulation: The Case of Responsive Pricing as a Substitute for Interruptible Power Contracts," *Marginal Costing and Pricing of Electrical Energy, A State-of-the-Art Conference*, Canadian Electrical Association, May, pp. 38–58.

Vierima, Teri L., and J. R. Malko (1981), "Natural Gas Rate Design: Innovative Activities in Wisconsin," *Public Utilities Fortnightly*, October 22.

Voorhees, Alan M., and Associates Inc., and Cambridge Systematics, Inc., (1976), *Carpool Incentives: Evaluation of Operational Experience*, Federal Energy Administration, Washington, D.C.

Walters, Frank S. (1978), "Alternatives to Marginal Cost Pricing in an Age of Energy Scarcity," *Marginal Costing and Pricing of Electrical Energy, A State-of-the-Art Conference*, Canadian Electrical Association, May, pp. 13–20.

Watkins, G. C. (1974), *Canadian Residential and Commercial Demand for Natural Gas*, University of Calgary Discussion Paper No. 30, May.

Waverman, L. (1977), "Estimating the Demand for Energy," *Energy Policy*, (March), 2–11.

Wilder, Ronald P., and John F. Willenborg (1975), "Residential Demand for Electricity: A Consumer Panel Approach," *Southern Economic Journal*, 42(2), 212–217.

Wildhorn, Sorrell, et al. (1976), *How to Save Gasoline—Public Policy Alternatives for the Automobile*, Rand Corporation, Santa Monica, Calif.

Wilson, John W. (1971), "Residential Demand for Electricity," *Quarterly Review of Economics and Business*, 11 (Spring), 7–22.

Wilson, J. W. (1974), "Electricity Consumption: Supply Requirements, Demand Elasticity and Rate Design," *American Journal of Agricultural Economics*, 56(2), 419–427.

Wisconsin, Department of Industry, Labor and Human Relations (DILHR) (undated), "Rules of Wisconsin Administrative Code, Chapter IND 18—Alternative Energy Tax Credits."

Wisconsin, Department of Industry, Labor and Human Relations (DILHR) (1978a), "Instructions to Apply for the Alternative Energy System Refund Claim for Individuals and the Income Tax Credit for Businesses/Corporations," December.

Wisconsin, Department of Industry, Labor and Human Relations (DILHR) (1978b), Uniform Dwelling Code, Chapter IND 22, Energy Conservation, December 1.

Wisconsin, Department of Industry, Labor and Human Relations (DILHR) (1978c), *Proposed Energy Conservation Standards for One- and Two-Family Dwellings, Environmental Impact Statement*, January.

Wisconsin, Department of Industry, Labor and Human Relations (DILHR) (1979), Safety and Buildings Division, "The Alternative Energy Income Tax Credit Program," Report to Assembly Committee of Revenue Hearings on AB 636, September.

Wisconsin, Office of State Planning and Energy (1978), *State Energy Conservation Plan*.

Wisconsin Power and Light Company (WP&L) (1978a), "Report on Load Management and Time-of-Day Rates," April 17.

Wisconsin Power and Light Company (WP&L) (1978b), "Solar Water Heating Fact Sheet."

Wisconsin Power and Light Company (WP&L) (1978c), "Solar Water Heating," brochure.

Wisconsin Power and Light Company (WP&L) (1978d), James G. Miller, "Time-of-Day Rates: Design, Implementation and Results," prepared for National Conference on the Impact of the National Energy Act on Utilities, Houston, December 5.

Wisconsin, Public Service Commission (PSC) (1977a), Docket No. 6690-ER-5, *Findings of Fact and Order Establishing Temporary Experimental Rates*, February 18.

Wisconsin, Public Service Commission (PSC) (1977b), Docket No. 05-GV-2, *Class A Gas Utility Residential Insulation Program, Findings of Fact and Order*, September.

Wisconsin, Public Service Commission (PSC) (1979a), Docket No. 58-UR-4, *Prepared Direct Testimony of Mo Reinbergs*, February.

Wisconsin, Public Service Commission (PSC) (1979b), Docket No. 05-GV-2, *Comprehensive Energy Conservation Programs of Class A Gas Utilities*, April 5.

Wisconsin, Public Service Commission (PSC) (1979c), Docket No. 01-ER-1, *Proposed Rule Regarding Peak Load Pricing/Time-of-Day Metering*, Findings of Fact and Order Dismissing Proceeding, May 24.

Wisconsin, Public Service Commission (PSC) (1979d), Docket No. 6690-UR-10, *Prepared Direct Gas Rate Testimony of Mo Reinbergs*, July.

Wisconsin, Public Service Commission (PSC) (1979e), Docket No. 6690-UR-10, *Testimony of Leigh A. Riddick*, July.

Wisconsin, Public Service Commission (PSC) (1979f), Docket No. 6690-UR-10, *Prepared Direct Electric Rate Testimony of Mo Reinbergs*, July.

Wisconsin, Public Service Commission (PSC) (1979g), Docket No. 05-GV-2, *The Comprehensive Plan and Report to the Public Service Commission*, submitted by Wisconsin Power and Light Company, September 28.

Wisconsin, Public Service Commission (PSC) (1980), *Supplemental Interim Order and Notice of Investigation Pertaining to Conservation in Rental Living Units*, December 2.

Wisconsin, Public Service Commission (1981), "Wisconsin Residential Time-of-Day Rate Demonstration Project," Office of the Chief Economist, April.

Wisconsin, Public Service Commission (PSC) (1982a), "Conservation Programs of the Public Service Commission of Wisconsin—A Summary from 1974–1981."

Wisconsin, Public Service Commission (PSC) (1982b), "Low-Income Weatherization: The Public Service Commission of Wisconsin Approach."

Wisconsin, Public Service Commission (PSC) (1982c), *Investigation on the Commission's Own Motion to Reconsider Whether Class A Private Electric and Gas Utilities Should Provide Conservation Financing to their Customers to Promote Weatherization and What Suggestions Can Be Made to the Commission for Improving Conservation Programs for Natural Gas and Electric Utilities*, Findings of Fact and Order, 05-UI-12, April 20.

Yang, Yung Y. (1978), "Temporal Stability of Residential Electricity Demand in the United States," *Southern Economic Journal, July*, 45(1), 107–115.

INDEX

315